TQM Simplified

TQM Simplified

A Practical Guide

James H. Saylor

Second Edition

McGraw-Hill

New York   San Francisco   Washington, D.C.   Auckland   Bogotá
Caracas   Lisbon   London   Madrid   Mexico City   Milan
Montreal   New Delhi   San Juan   Singapore
Sydney   Tokyo   Toronto

Library of Congress Cataloging-in-Publication Data

Saylor, James H.
    TQM simplified : a practical guide / James H. Saylor.—2nd ed.
       p.   cm.
    Rev. ed. of: TQM field manual. 1st ed. c1992.
    ISBN 0-07-057678-5
    1. Total quality management. I. Saylor, James H.
TQM field manual. II. Title.
HD62.15.S29 1996
658.5'62—dc20   96-6192
                                                    CIP

McGraw-Hill

A Division of The McGraw·Hill Companies

The first edition of this book was published copyright 1992 under the title *TQM Field Manual*.

1 2 3 4 5 6 7 8 9 0 DOC/DOC 9 0 1 0 9 8 7 6

ISBN 0-07-057678-5

The sponsoring editor for this book was Larry Hager, the editing supervisor was Fred Bernardi, and the production supervisor was Don Schmidt. It was set in Century Schoolbook by Ron Painter of McGraw-Hill's Professional Book Group composition unit.

Printed and bound by R. R. Donnelley & Sons Company.

This book is printed on acid-free paper.

To James, Jr., Joe, and Christina, my children; Tyler my first grandchild; and all future generations of children hoping for a world of peace and prosperity.

Contents

Preface

"America, you lost the economic war!"
—Koito Manufacturing,
Annual Meeting (June 1990)

This quote introduced the original *TQM Field Manual*. This economic war continues now, more than ever, on a global front and it will never end. It is a war that's always in its final campaign, and as such, it will require the continuous achievement of victories. America has won some of its battles in this economic war, but it also continues to suffer some casualties and much destruction. Every day we see the evidence of this loss with unemployment, homelessness, and crime. America keeps experiencing major destruction of its infrastructure. Education, government services, and even family values continue to deteriorate. As we look forward, Americans must face the many challenges of achieving victories in this economic war.

Victories in this economic war are just as important as victories in any military war. In fact, the impact can be even greater. Survival as an economic power is at stake. There are some differences in an economic war that make it difficult to achieve victory. First, VICTORY cannot be clearly defined for all. VICTORY varies for each specific organization. Second, VICTORY is forever changing. Third, the enemy is not always evident. The enemy can be any element that keeps the organization from getting and keeping customers. Enemy elements can be competition, technology, government—even ourselves. Fourth, the war never ends. Although the organization may achieve many victories, ultimate VICTORY is a never-ending process in an economic war. To strive to achieve VICTORY in this war, the organization must learn to adapt to today's environment with a view toward tomorrow.

To achieve VICTORY in an economic war, a country, a company, or an organization must have customers. This endeavor to get and keep customers requires the unrelenting pursuit of customer satisfaction, which must be the focus for all Total Quality Management (TQM) efforts. TQM, itself, should never be the ultimate focus. TQM is only the means to achieve the ends or VICTORY—the VICTORY as defined by each specific organization.

Although VICTORY can be different for every organization, there is no difference in the aim of TQM for an individual; an entrepreneur; a commercial enterprise; a small, medium, or large business; a nonprofit association; or a government agency striving to achieve victory. There is no difference between a public or private operation. There is no difference in a service- or product-oriented organization. The basic focus is always to satisfy customers' needs.

As individuals and organizations become engaged in the struggle to survive they get caught up in the day-to-day operations, the fire-fights, the endless crises, and the temptation to seek short-term profit no matter the consequences. We look for the quick fix. We look at TQM, quality, teams, concurrent engineering, reengineering, open management, and so on, as ends in themselves. We sometimes lose the focus that customers are the key to long-term survival and prosperity.

Theodore Levitt of the Harvard Business School in his book *Thinking about Management* (Free Press, 1991) states it best:

> "Quality" becomes an imposed dictum rather than an understood dedication.
> "Fast response" a mechanical metric rather than a meaningful motivation.
> "Market share" a warlike target rather than an earned result.
> "Self-managed teams" a comfortable indulgence rather than a purposeful construct.
> "Entrepreneurship" an escape from discipline rather than a leap into enterprise.
> "Restructuring" a game of finance rather than a fight for market effectiveness.

The total systematic, integrated, consistent, organizationwide approach to Total Quality Management is advocated as the approach to achieve VICTORY. This perspective looks at TQM as a total, integrated system instead of emphasizing one aspect as most important.

The total, integrated systems approach that stresses the human element is essential for success in the future. People are the key to success in all improvement efforts. When all the models, methodology, tools, and techniques are examined, the end result is that people make it happen. People are the most valuable resource in any TQM effort. It is the combination of basic technical tools and techniques of

TQM with the emphasis on the importance of the human element that makes success.

The war analogy is used in this preface to convey the seriousness of the situation. We are engaged in a total economic war. Our very survival as an economic force is at stake. Already there have been many casualties. Organizations and people have been destroyed. Many organizations and people have been wounded. It impacts all aspects of our economy. It affects both private and public organizations.

The economic war is always in the final campaign. In order to achieve victories in this, the final campaign, many changes from traditional methods are required. Many constantly changing factors can contribute to victories at any one time. This effort requires a continuous reexamination and revamping of an organization's systems and processes. Total Quality Management is the process used to turn defeat into victory. Both industry and government must use Total Quality Management; America's future depends on our success in this war.

Remember, this is a war. Wage this war with the same fervor and support as any other war. Survival depends on it. Everyone must work toward VICTORY.

James H. Saylor

Acknowledgments

A special acknowledgment to all the people who helped with the *TQM Field Manual* and this book: Ellen Domb, Peter Angiola, Louis Chertkow, Don Wolking, Gene Leist, Pete Weber, Dick Rawcliffe, Jack Duffy, Dave Ford, Brandon Hamilton, Frank Herrmann, Jeff Hill, Stewart Horwitz, Norman Michaud, Rosann Saylor, and Gordon Tyler.

Also, thanks goes to customers of The Business Coach for giving me the chance to experience all aspects of the Total Quality Management efforts. Special thanks to Aerojet Electronics Systems Division, GenCorp Aerojet, Elkay Plastics, and the Society of Logistics Engineers in the U.S. and Europe.

In addition, special acknowledgment to Bruce T. Barkley, my coauthor in *Customer-Driven Project Management*.

Recognition goes to Rosann D. Saylor whose dedication and continuous assistance in many aspects of this book greatly contributes to its overall quality.

As with all worthwhile efforts today, this was truly a "team" effort. All members of the team deserve special recognition for their specific contribution to customer satisfaction.

Introduction

TQM Simplified provides the basic core of the leadership and management process focusing on meeting the many challenges facing organizations as we look forward to the next century. TQM is a difficult journey that takes years in many organizations; this book simplifies this complex process. It provides simple, practical proven processes, methods, tools, and techniques to make the trip as smooth and easy as possible for everyone from the top executive, to middle managers, to workers.

The focus of this book is the application of Total Quality Management to help organizations succeed in achieving the results needed for their specific VICTORY today and in the future. It is a "how-to" book for continuous improvement of processes, product/service, and people. It helps any organization target appropriate business results, through continuous improvement of the business processes. This effort produces big gains impacting productivity, product/service quality, costs, and cycle time.

This guide targets the creation and maintenance of a Total Quality Management system to use as a means to achieve VICTORY—the VICTORY as defined by each specific organization. You will get essential, valuable, practical information that you can apply immediately to make improvements to your organization. This book broadens and updates the VICTORY-C model introduced in the *TQM Field Manual*. It simplifies the basic TQM philosophy, guiding principles, approaches, continuous improvement systems, improvement methodologies, tools, and techniques.

The aim of this book is to provide specific "how-to" guidance for any organization to achieve its specific VICTORY. This goal is accomplished through step-by-step action processes which guide any practitioner to results.

TQM Is Not Simple

TQM simplified does not mean to imply that Total Quality Management is simple. Although the "what" of TQM is easy to understand and readily embraced as the approach for success, the "how to" of TQM is complex and involves the application of skills in many disciplines. The major reasons that a TQM process fails can be traced mostly to misapplication. In fact, the purpose of this book is to take the highly complex and misunderstood approach to management of an organization and break it down to its basic processes, methods, tools, and techniques that can be helpful to any organization. This book provides the basic TQM action processes that anyone in the organization can apply immediately in any organization.

TQM Is Not Only About "Product or Service Quality"

TQM goes beyond basic product or service quality. Product and service quality are minimum requirements for any organization in today's world. TQM is about organizational success or VICTORY. This book encompasses all the elements of TQM. "Total" includes everyone and everything. "Quality" equates to total customer satisfaction. "Management" means the organization's leadership and management approach.

TQM Is Not a Program

TQM is a way of organizational life for survival and growth. TQM is a never-ending process. It is a systematic, organizationalwide, integrated process for operating an organization. In many organizations today, the emphasis is not on specific programs for "quality" but on the achievement of "customer satisfaction" in a cost-effective way as the manner of doing business.

TQM Is Not Just a Management Fad

Plainly put, TQM is the "best business practices for the times." TQM has proven itself beyond the usual six months of most management fads. TQM has become a proven leadership and management approach since the middle 1980s. The term TQM is still widely used by many organizations. In other organizations, they are just doing TQM as normal course of day-to-day operations.

TQM Is Not an End in Itself; TQM Is Only a Means to an Organizational End

TQM should never be the focus of a business. TQM is only the means to achieve the focus of the business, namely, to satisfy the customer, which it needs to exist. TQM focuses an organization toward the customer. In addition, for any for-profit organization, TQM is the means to this goal. For example, the traditional formula for profit is as follows: revenue − cost = profit. I like to rewrite this formula with the right emphasis: customers + competitive processes = profit. This puts the focus in the proper order. First, customers provide revenues. TQM targets getting and keeping customers. The amount of profit an organization gains also depends on costs, as stated in the traditional formula. TQM also addresses the management of cost. Competitive processes give the organization the ability to satisfy customers in the most efficient and effective manner; thus, costs can be managed. Notice, cost management, not cost reduction, is the emphasis of TQM. Again, cost reduction where appropriate is a normal consequence of cost management. However, cost management implies intelligent use of funds so that the organization can develop, grow, and beat the competition.

TQM simplified provides

- A simple, easy-to-use, "how-to" desktop reference

- A systematic, integrated, consistent, organizationwide approach

- Easy-to-remember processes for achieving results

- A book for everyone yearning for "real" results

Since every organization is different, this book provides processes that can and are intended to be supplemented with content information from each organization. You are encouraged to tailor the processes to your specific situation to assist your performance. You use the processes from this book and your specific content knowledge of your situation to achieve your victories.

There are some cautions regarding tailoring. First, you must never violate TQM philosophy or any of its guiding principles. If you pick and choose only certain parts of the TQM philosophy or guiding principles, complete VICTORY will never be achieved. Second, you must establish and maintain all the elements of VICTORY-C. If you violate either of these two guidelines, not only will VICTORY elude you, but defeat is guaranteed.

To strive to achieve success, the organization must learn to adapt to today's environment with a view toward tomorrow. Currently, there are many approaches to meet this challenge. These approaches have various names, for example: Total Quality Control, Total Quality Improvement, Total Improvement, Total Quality, Quality Leadership, Quality Improvement, Process Improvement, Continuous Measurable Improvement, Employee Involvement, and more. These approaches can be considered as Total Quality Management (TQM). TQM is not one specific "best" approach. Rather, it is a term that has evolved to encompass all efforts to achieve VICTORY. TQM is the philosophy and set of guiding principles required by any organization to strive for success.

This book provides a guide for everyone involved in the Total Quality Management process or anyone just wanting to do better. The basic framework along with some in-depth information is provided to allow everyone to determine the specific elements necessary for them to achieve success. This book can be read cover-to-cover, or specific sections can be skimmed for particular information. It can also be used as a "how-to" book for certain TQM methodologies. Further, it can be used as an education and training book for an introduction to all aspects of Total Quality Management.

I hope this book helps you to achieve your specific VICTORY.

1

The Basics of Total Quality Management

Total Quality Management is a leadership and management philosophy and guiding principles stressing continuous improvement through people involvement and quantitative methods focusing on total customer satisfaction.

Many organizations use a Total Quality Management (TQM) approach to achieve supreme excellence. TQM is a process aimed at transforming an organization into one capable of achieving success in an ever-changing environment. TQM provides any organization the means to meet the many challenges of today while ultimately moving the organization toward the future. TQM focuses the organization on continuous improvement geared to total customer satisfaction. This customer-oriented process combines fundamental management practices with existing improvement efforts and proven tools and techniques. TQM is applicable to every organization striving to be the best, whether that organization is one function, a division, an operating agency, a company, or a corporation. Total Quality Management is equally useful for large and small businesses, manufacturing and service industries, and public and private organizations.

TQM Basic Considerations

Total means the involvement of everyone and everything in the organization in a continuous improvement effort. This not only includes all the people but it also encompasses all the systems, processes, operations, and equipment.

Quality is total customer satisfaction. Total customer satisfaction is the center or focus of TQM. The customer is everyone affected by the product and/or service and is defined in two ways. The customer can be the ultimate user of the product and/or service, known as an external customer. Or the customer can be the next process in the organization, known as an internal customer. TQM focuses on satisfying all customers, both internal and external.

Management refers to people and processes. First, management is the leaders of an organization. Management creates and maintains the TQM environment through leadership and empowerment. Further, the management of the organization ensures quality (customer satisfaction) by continuous improvement of processes, products and/or services, and people. Second, management refers to the process of planning, organizing, staffing, directing, and controlling. These functions are all critical to a TQM organization.

TQM Background

TQM evolved over many decades to meet today's challenges.

A full exploration of the concept of Total Quality Management requires an examination of what has happened over the years to change some of the traditional theories of management and the classic approaches to quality. Total Quality Management has evolved from a diverse set of many "traditional" management and quality ideas.

Historically, management concepts focused on the functions of organizational control, including planning, organizing, staffing, directing, tasking, structuring, coordinating, budgeting, evaluating, inspecting, and reporting. These concepts were primarily derived with industrial and corporate firms as their models. However, many other organizations were influenced by these theories, including government and nonprofit organizations. This increasing influence led to the wide acceptance of these management principles. The basic assumptions were that managers are paid to exercise control of the organization's resources, including people, in order to ensure that they meet the objectives of management of the organization. Scientific principles such as unity of command, span of control, uniformity, centralization, delegation, discipline, and work flow were emphasized. This scientific approach to management was derived and promoted by Frederick Taylor, Henry Fayol, and Max Weber to support the concept of efficiency and control. Production was more important than people, and people were "slotted" into job functions. The management of things was the emphasis. In this environment people were specialized or eliminated. The output was the major focus.

Psychologists, including Lillean Gilbreath, following on abuses of the industrial revolution, began to focus on the human element in the workplace. They concluded that it was not the monotony of work that caused employee dissatisfaction, rather it was management's lack of interest in the workers. Then Elton Mayo defined the organization as a social system and concluded that work groups determine attitudes and behaviors, not top management, and that productivity is directly related to the satisfaction of the employees and particularly to the amount of attention paid to them (the "Hawthorne effect"). Many others then tried to integrate productivity and the human elements. They included Peter Drucker through management by objectives, and Blake and Mouton through the managerial grid. Douglas MacGregor defined two types of managers: theory X managers, authoritarians, who believe that people are inherently lazy and do not want to perform, and theory Y managers, nurturers, who believe that people are innately responsible and capable of exercising initiative and making worthwhile contributions. Other behaviorists worked to explain motivation in the workplace.

As the "traditional" management approaches transformed since the turn of the century, quality techniques evolved after World War I. The concepts which were to crystallize into Total Quality Management began with the statistical quality concepts of Walter Shewhart as published in 1931 in his book *Economic Control of Quality of Manufactured Product*. His statistical approach called statistical process control is the foundation for the "quality" management approach. His approach was refined by many practitioners during World War II to improve the quality and productivity of America's war products.

This quality approach has evolved continuously to today's Total Quality Management. Total Quality Management includes a wide range of management practices, methods, tools, and techniques. Since TQM is a collection of the best of many disciplines focusing on the processes required to survive and prosper in the environment of global competition, there are many contributors. Some of the major contributors include W. Edwards Deming, Joseph M. Juran, Armand V. Feigenbaum, Kaoru Ishikawa, Genichi Taguchi, Philip B. Crosby, Peter Drucker, Tom Peters, H. James Harrington, A. Richard Shores, and many others in government organizations, most notably the U.S. Department of Defense.

TQM has its foundation in the "quality" movement . The Total Quality Management approach evolved to meet the needs of today's global environment, the result of world economics since World War II. After World War II, in an American industrial world that had little competition from foreign manufacturers and service providers, quality was not seen as very important. Everything that was made was sold, almost regardless of quality, since America was in most cases the only producer. Quality in the post-World War II years continued to be a second-class citizen. First priority was placed on quantity and production, getting the products out the door. Companies employed inspectors at the end of the process to find defects and rework them.

At the end of the World War II, America was the leading producer in the world, and American industrial leaders did not feel the need to continue the quality push. However, Japan demanded an economic rebirth. Japan viewed quality as an essential catalyst to their economy. Japan sought the assistance of many of America's quality experts.

This next stage in the quality movement was stimulated by Japan with the assistance of American quality experts. One of these American experts, W. Edwards Deming, helped the Japanese focus on their quality obsession. The primary motivation for the quality vision in Japan was the creation of jobs. The Japanese determined that to recover from the war, they had to transform their industries to enable them to produce quality commercial products. Deming showed the Japanese how they could improve quality and productivity through statistical techniques to capture more business and create

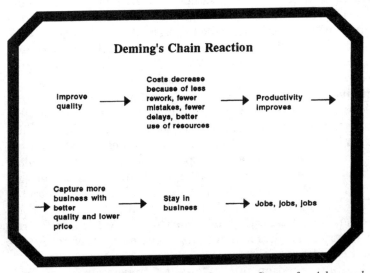

Figure 1.1 Deming's chain reaction. (*Courtesy Center for Advanced Engineering Study*)

jobs. Figure 1.1, from Deming's book *Out of Crisis,* illustrates the Deming chain reaction that played a major role in Japan's focus on quality. Simply, quality improvement became the vision for everyone in Japan.

W. Edwards Deming helped the Japanese with their obsession with quality. His 14-point approach to quality was originally detailed in his book *Out of the Crisis.* (This version was provided courtesy of W. Edwards Deming through the Massachusetts Institute of Technology, Center of Advanced Engineering Study, January 1991.) His updated 14 points are as follows:

1. Create and publish to all employees a statement of the aims and purposes of the company or other organization. The management must demonstrate constantly their commitment to this statement.

2. Learn the new philosophy, top management and everybody.

3. Understand the purpose of inspection, for improvement of processes and reduction of cost.

4. End the practice of awarding business on the basis of price tag alone.

5. Improve constantly and forever the system of production and service.

6. Institute training.

7. Teach and institute leadership.

8. Drive out fear. Create trust. Create a climate for innovation.

9. Optimize toward the aims and purposes of the company the efforts of teams, groups, staff areas.

10. Eliminate exhortations for the work force.

11. (a) Eliminate numerical quotas for production. Instead, learn and institute methods for improvement.

11. (b) Eliminate M. B. O. (Management by Objectives). Instead, learn the capabilities of processes, and how to improve them.

12. Remove barriers that rob people of pride of workmanship.

13. Encourage education and self-improvement for everyone.

14. Take action to accomplish the transformation.

There were many others who also assisted the Japanese in pursuing their quality vision during the next decades after World War II. The most notable were Joseph M. Juran, Armand V. Feigenbaum, Kaoru Ishikawa, and Genichi Taguchi. From these early gurus, many others were fostered and continuously improved to make Japan an economic world power.

Joseph M. Juran, a leading quality planning advocate, was another American, like Deming, instrumental in Japan's early success. He taught the Japanese his concepts of quality planning. Both Juran and Deming stress traditional management as the "root" cause of quality and productivity issues. Juran focuses on a disciplined planning approach to quality improvement. He published his *Quality Control Handbook,* including the application of basic statistical control to business practice. The message was that our organizations are smothering people in command-and-control systems and that the workers do not feel empowered to think and act on the basis of their insights into how business is practiced and customers served.

The Juran approach to quality is based around three major processes, "the Juran trilogy," which consist of quality planning (design for quality), quality control, and quality improvement.

Armand V. Feigenbaum, also an American, was the first to use the term "Total Quality." His book, *Total Quality Control,* is one of the best early works on Total Quality Improvement. It is now in its third edition, published by McGraw-Hill in 1991, and is available separately in a special Fortieth Anniversary Edition. His quality improvement approach is a systematic, integrated, organizationwide perspective. He also originated the concept of the cost of quality, which monitored cost of failures, quality appraisal, and prevention costs. This approach aimed managers toward quality improvement through a reduction in quality costs.

Kaoru Ishikawa, Japan's leading expert, geared the quality vision to the masses. He stressed the seven basic tools of quality used for problem solving. He believed that almost all quality problems could be solved using the seven basic tools. These tools include Pareto charts, cause-and-effect diagrams, stratification, checksheets, histograms, scatter diagrams, and control charts.

Genichi Taguchi, another one of Japan's top quality experts, was the first to stress proper design strategy. He redefined the concept of design specifications. Simply being within specifications is not good enough. He introduced a methodology that focuses on optimizing the design. According to Taguchi, any variation of performance from best target values is a loss. Loss is the enemy of quality. The goal is to minimize loss by focusing on the best target value.

In general, the Japanese adapted, developed, and continuously improved the quality approaches of these early pioneers. They formulated the concept of continuous improvement. As a result of the continuous improvement philosophy, they advanced through many small innovations several products originally developed by the United States, including televisions, automobiles, cameras, small electronics, air conditioners, video- and audiotape recorders, telephones, and semiconductors. The list is endless.

The next stage of the quality movement started in the United States in the late 1970s. After World War II until this time the United States did not feel the need to embrace the quality vision. America was the number-one economic power in the world, and the world bought whatever the United States produced. During the late 1970s, the threat of competition from other countries became apparent to many American industries. These industries started to investigate ways to become more competitive. This effort resulted in a renewed interest in the quality management techniques being used by the competition—mainly Japan.

During this stage, the American quality movement was reborn. Many of the early quality experts' teachings were updated. In addition, many others joined the march to quality and provided additional insights into a transformation process for America. This new movement in America goes beyond the quest for quality. It calls for a transformation of America's management. This new management philosophy evolved into Total Quality Management. American TQM strives to develop an integrated system that takes advantage of America's strengths, particularly its people resources. America's strength lies in its people's diversity, individuality, innovation, and creativity. These resources, especially the people resources, should be targeted and maximized on quality. The ultimate goal is total customer satisfaction from every customer, both internal in and external to the organization. The quality obsession in America targets customer satisfaction. To focus the entire organization on total customer satisfaction requires management to create a TQM organizational environment. This TQM work environment where everyone can contribute with pride is developed through leadership. As it exists today, this American style of TQM stresses a totally integrated, systematic, organizationwide perspective. It requires a transformation of many of the ways America traditionally does business.

This transformation process has been evolving since the late 1970s through the efforts of many American organizations. Some of the early leaders were organizations that realized that their survival in the new global economy required some fundamental changes. These were organizations like Xerox, Motorola, IBM, and Hewlett-Packard. By necessity, they quickly learned the lessons of continuous improvement. In the government area, the Department of Defense is one of the leaders in Total Quality Management. In the mid-1980s, the

Department of Defense was facing an ever-declining budget. They sought assistance from the quality experts to help them determine how they could protect America at a lower cost. In addition to the many organizations, there are many people who contributed to refining the new American management movement. Besides using concepts of quality masters like Deming, Juran, Feigenbaum, Ishikawa, and Taguchi, TQM embodies the ideas of many others from quality, management, organizational development, training, engineering, and other disciplines. The ever-evolving list of contributors is too numerous to mention. Some of the better known early contributors include Philip B. Crosby, Tom Peters, Robert H. Waterman, Jr., H. James Harrington, and A. Richard Shores.

An early proponent in the late 1970s, Philip B. Crosby outlined the Zero Defects system in his book *Quality is Free*. His program was based on many years with Martin Marietta and ITT. It was embraced by many American companies and the U.S. government. The program was instrumental in uncovering many defects evident in the American industrial process. The Crosby approach is based on four points: (1) quality is conformance to requirements; (2) prevention is the key to quality; (3) Zero Defects is the standard; and (4) measurement is the price of nonconformance.

In the early 1980s, the media and popular literature gave more attention to quality. Tom Peters and Robert H. Waterman, Jr., provided American business with more ideas on what contributed to the success of the top companies in America. Their book *In Search of Excellence* presented an initial, inside look at what made these companies so competitive. They determined that the following eight attributes distinguish excellent, innovative companies. First, these organizations are geared for action. They prefer to do something rather than going through endless analysis and committee reports. Second, they continuously strive to meet the needs and expectations of their customers. Third, innovative organizations are structured with smaller organizations within, allowing internal autonomy and entrepreneurship. Fourth, successful organizations foster the ability to increase productivity through people. Fifth, they are value driven with management setting the example through the application of hands-on attention to the organization's central purpose. Sixth, they build on organizational strength by sticking to what they do best. Seventh, successful organizations have few layers of management and few people in each layer. Eighth, excellent organizations create an atmosphere of dedication to the primary values of the company and a tolerance for all employees who accept those values. These eight attributes still characterize many capable organizations of today.

In 1987, Tom Peters, in *Thriving on Chaos,* proclaimed there are no excellent companies. He attributed this new view to the constant changes in the new environment. A company in the new environment must continuously improve or other companies will replace them. Therefore, a company cannot ever achieve excellence. Peters called for a management revolution in America—a revolution that is constantly adapting to the many challenges in the economic environment of today. Tom Peters details a total management transformation process with prescriptions for proactively dealing with today's environment.

In the late 1980s and early 1990s, many other Americans updated the teachings of the earlier quality experts to the environment of today. American approaches grew from various applications in many different large companies and government organizations. Again, the many contributors are too numerous to mention. For proven approaches, H. James Harrington from IBM, A. Richard Shores from Hewlett-Packard, and James H. Saylor from GenCorp Aerojet provide excellent, proven road maps based on practical experience for Total Quality Management in American industry today. Harrington's approach is described in his books *The Improvement Process* and *Business Process Improvement.* Shores' book *Survival of the Fittest* outlines his teachings. Saylor's approach was first outlined in the *TQM Field Manual.* The Harrington approach focuses on the entire improvement process. Shores stresses Total Quality Control. The *TQM Field Manual* introduced the VICTORY-C approach that emphasizes a systematic, organizationwide, consistent effort. These quality practitioners provide many specific methods geared to today's American industry.

In 1987, the need for quality improvement was formally recognized by many industry leaders and the U.S. government with the creation of the Malcolm Baldrige National Quality Award. As part of a national quality improvement campaign, the Malcolm Baldrige National Quality Award was created by public law to foster quality improvement efforts in the United States. The annual award recognizes U.S. companies in the categories of manufacturing, service, and small business that excel in quality achievement and quality management. This award's criteria for leadership, information and analysis, strategic quality planning, human resource utilization, quality assurance of products and services, quality results, and customer satisfaction has been continuously improved since its inception. The award is recognized as the standard for all organizations in the United States trying to pursue excellence.

In the 1980s the federal government began its quest for quality and productivity improvement. In the early 1980s, some defense logistics organizations began exploring methods to enhance their performance. This initiative resulted in the eventual spread of systematic improvement efforts to other Department of Defense (DoD) organizations throughout the decade. In 1988, the DoD adopted the TQM approach. TQM was to be the vehicle for attaining continuous quality improvement within DoD and its many DoD contractors.

Also in the 1980s, other federal agencies initiated productivity and/or quality improvement ventures. These agencies included the Internal Revenue Service, NASA, General Services Administration, and the Departments of Agriculture, Commerce, Energy, Interior, and so on. By mid-1988, Total Quality Management had evolved into a governmentwide effort. This evolution was formalized by the establishment of the Federal Quality Institute as a primary source of information, training, and TQM services. In the late 1980s, a President's Award for Quality and Productivity Improvement was created to recognize the agency or major component of an agency that has implemented Total Quality Management (TQM) in an exemplary manner, and is providing high quality service to its customers. This award, like the Malcolm Baldrige Quality Award, formally established the commitment to quality in the federal government.

In the 1990s, Total Quality Management is being adapted in many other government agencies, communities, and private industries in the United States under many names such as Total Quality Leadership (TQL), Total Quality Improvement (TQI), Continuous Quality Improvement (CQI), total customer satisfaction, and so on. In particular, educational institutions and state, county, and city government agencies are discovering that change is necessary. For instance, many higher education institutions are implementing Total Quality Management. Small businesses are discovering TQM. In recent years, health-care organizations are starting on the road to continuous quality improvement. They are finally understanding: America can procrastinate no longer. The "sense" of this movement is that all of America must be transformed to restore its position as a leader in the world economy.

Today, Total Quality Management is established as a proven approach to success in today's world with a view for the future for organizations large and small, public or private, profit or nonprofit, product or service oriented.

Why Total Quality Management?

TQM is the means to achieve an organization's end results.

Total Quality Management meets the challenges of a global economic environment. It provides a management approach adaptable to the
new world of rapid change, rising complexity, and rabid competition.
Today, political, technological, social, and economic changes are swift.
In the short period since the end of the 1980s, the world has been
turned around. For instance, the U.S.S.R.'s empire has crumbled. The
Berlin Wall has been torn down. Japan has become the world's number-one economic power. The United States is just one of many players
in the global marketplace. With technological advances, especially in
computers and telecommunication, the information age has launched
a time of increasing intricacy in all of the world. This development has
brought about rising complexities in the processes used to perform our
work. Competition on a global scale is a fact of life. Everyone is competing for the new global markets. With competition fierce in all aspects—technology, cost, product quality, and service quality—everyone
must seek new approaches to be competitive and share in global economic growth. Total Quality Management provides an approach that
can be used to confront these challenges today and in the future.

Today's world

> The world is not the same today as yesterday.

Today's world is radically different from the recent past. It is a new
environment where old solutions no longer work. The "same old way"
simply does not bring about the necessary results. Technology is not
the prospective cure-all. Throwing resources at the project for short-
term progress does not foster long-term customer satisfaction. Our
paradigm (mindset) must change to reflect the reality of today's world
for us to achieve success.

Some of the major considerations in today's world are shown below.
These conditions demand change. The following current organizational issues may be affecting your organization. Check the items that you
think are applicable to your situation. If your organization is facing
one or more of these considerations, you should consider using TQM:

- Competitive environment
- Uncertainty in the organizational environment
- Need to work smarter
- Changes in management and leadership philosophy, principles,
  methods, tools, and techniques

- More selective customers who increasingly demand added value
- Conservation of limited resources
- External factors affecting the organization which are progressively out of control
- The organization's systems constantly requiring updating to optimize productivity, quality, and costs
- New or changed products and services perpetually being introduced
- Continuous vigilance needed of all factors affecting the organization, the product and/or service, the competition, and the customers
- Economic pressures causing requirement for cost management
- Shrinking budgets
- Trying to stress, more than ever, "more for the buck"
- Rapidly changing technology making stability difficult
- Accelerated production time being essential to a competitive advantage
- Customer-driven quality being critical to long-term growth
- Application of telecommunication and information systems

There are many players in the competitive global economy. Japan has replaced America as the new world economic leader. In addition, there are many other formidable players. This fresh economic playing field requires everyone to transform their management philosophy, principles, methods, tools, and techniques into a management system that allows everyone to work smarter to rapidly respond and satisfy customers.

Rabid competition is the way of the world economy. Just because a product or service is available does not mean it will sell. The customer is more selective in buying goods and services. In fact, as more and more options become available, the customer becomes increasingly discriminating, demanding added value. This fact makes keeping and getting new customers more important than ever. Customer satisfaction is the focus of all competing organizations. The organizations that can answer constantly changing customers' demands will succeed in this new environment of rabid competition.

Uncertainty is now always a concern. With a rapidly changing world order, certainty can no longer be taken for granted. No organization is safe from some sort of distress. External factors affecting the organization are progressively out of control. The organization's systems are constantly requiring updating to optimize productivity, quality, and costs. New or changed products and services are perpetually being introduced. Stronger competition is increasingly the norm. In addition, customer needs are continuously changing. Continuous vigilance of all factors affecting the organization, the product and/or service, the competition, and the customers is a necessity.

Economic pressures are a fact, which makes cost and budget a factor today and tomorrow. Lowest possible cost is the aim of all internal processes. It is no longer good enough to strive for reasonable cost. Everyone has the same technological advantages to make use of economies of scale, automation, and other production and service techniques to reduce costs. Customer satisfaction and profits in today's world depend on providing a product and/or service at the optimum, lowest possible cost.

In addition, economic pressure makes optimizing budgets an everyday reality. Currently, budgets are shrinking in most organizations, which causes a reexamination of priorities to stress more than ever "more for the buck." Economic pressures will continue to dominate choices and decisions in public and private organizations. The demand for increased value at less cost will continue into the next century.

Rapidly changing technology makes stability impossible. Failing to keep pace with the latest technology can bring obsolescence within a short period of time. Many products today have a very short life cycle. The impact of new technology, especially in information processing and communications, may determine supremacy.

Conservation of limited resources is a necessity. Global competition for scarce resources will only increase in the new global marketplace. Compounding competition, however, is the need to protect and preserve the environment. With many nations competing for few resources, coupled with the concern for the environment, waste and loss are everyone's enemies. Organizations must learn new techniques of quality, productivity, and project delivery focusing on elimination of variation to optimize all resources.

Reasonable production times no longer meet customers' needs. Accelerated production time is essential in many industries. The organization that is first to the marketplace is usually the winner. In today's world, speed is a competitive advantage.

Customer-driven quality is critical to long-term growth. Since the customer defines quality by their satisfaction, the supplier forcing a deliverable on a customer does not foster customer satisfaction. Customer satisfaction pulls quality from the supplier. Today, the customer or customer's voice must direct every aspect of the deliverable. This method is the only way to ensure quality.

Today's world is telecommunication- and information-controlled. Gone are the days when hardcopies, space, and distance were dominating factors of competition. The organization that can speed the right information to the right place at the right time is ahead of its rivals.

Today's world demands change

> *Successful organizations are changing to meet the new challenges.*

To adapt to today's economic world with an eye to the future requires the organization to be totally responsive to the customer. Specifically, the successful organization will be the one that can change to apply the new paradigm for prosperity in today's global environment. The following is a list of changes some organizations are making for the future. Each organization needs to determine the items that are applicable to their specific situation.

- Focus for organization
- Systems view
- Continuous improvement of quality and productivity
- Concurrent engineering
- Quality deliverables
- Constant improvement, reengineering, and inventing of internal processes
- Flexible organizational structure
- Few organizational layers
- Cooperation while competing
- People who add value
- Strong management and leadership
- Long-term view
- New performance reward systems
- Education and training for everyone
- Total customer satisfaction being the driving force

Table 1.1 lists some of the changes demanded in today's world.

Continuous improvement of processes, people, and products aimed at customer satisfaction is essential. The "if it's not broke, don't fix it" attitude does not promote critical thinking necessary for growth. Continuous improvement is the only way to survive. This attitude is the proactive approach to change. This new view of "everything can

TABLE 1.1   The World Demands Change

From	To
If it's not broke, don't fix it	Continuous improvement
Functional orientation	Systems view
Sequential design and production	Concurrent design and production
Inspection	Prevention
Quality not important	Quality critical
Accept current processes	Improve, reengineer, invent processes
Development	Innovation
Rigid organizational structure	Flexible organizational structure
Many organizational layers	Few organizational layers
Compete	Cooperate to compete
Individual performance	Team performance
People specialized, controlled, eliminated	People adding value, flexible, adaptable, empowered
Strong management	Strong management and leadership
Leadership only at top	Leadership everywhere in the organization
Short-term outlook	Long-term vision
Individual merit reward system	Team performance reward system
Education for management	Education and training for everyone
Driven by profit	Focus on total customer satisfaction

be made better through process improvement" stimulates the creativity and innovation needed to constantly grow. In addition to continuous improvement of processes, there must be a continuous improvement of people through a constant upgrading of their knowledge and skills by means of a lifelong learning system. Further, products and/or services require progressive enhancement to meet or exceed changing customer needs and expectations.

Systems thinking must replace functional orientation. In today's world, everyone's horizon must be expanded beyond narrow-minded occupational disciplines like engineering, manufacturing, accounting, education, training, logistics, and so on. Organizations struggling for success in the world economic environment cannot afford to subsidize functional "fiefdoms" which suboptimize resources. Progressive organizations must view the combination of all their processes as a system focused on customer satisfaction. This orientation requires everyone in the organization to have a systems outlook geared to achieving organizationwide excellence. This goal can be achieved by combining quality and performance.

Concurrent design is necessary, especially in most industries where time-to-market is a competitive advantage. Time-to-market is increasingly a differentiator in the marketplace for both products and services. Concurrent design of products and services reduces time-to-market significantly over traditional sequential design methods.

Inspection-based quality assurance needs to be supplanted with prevention techniques. Again, the industrial mindset expects defects. This inspection-based viewpoint adds excessive cost to the product or service. This is a cost that most organizations can no longer afford and the customer does not need to support. By shifting the emphasis to prevention techniques, the right thing is done right the first time. This effort reduces cost while increasing product and/or service quality. Prevention techniques focus on the improvement of all the processes in an organization to maximize the capabilities of processes.

Quality focused on customer satisfaction is required. Disregard for quality or just meeting conformance to requirements is no longer good enough to keep and maintain customers. Today's knowledgeable customers demand satisfaction, which is how they define quality. So the organization must determine what will satisfy the customer, and then focus the whole organization on striving to meet the customer's needs and expectations. To do so, the organization must identify the elements of quality that are of vital importance to its particular customers. The key elements of quality for a customer might include the following items: perceived product and/or service quality, performance, reliability, supportability, durability, features, availability, aesthetics, serviceability, maintainability, usability, environmental impact, conformance to requirements, customer service, logistics, training, warranty, and life-cycle cost. For instance, some customers demand reliability as a key element of customer satisfaction. In other cases, the customers might value performance as the critical element. For other customers, the key element of quality is availability. For them, it is of primary importance that the product or service be available in the right quantity, at the right time, and at the right place.

Innovation through constant incremental improvement must be pursued. Innovation rather than major development is the key for many organizations and industries. Building on the old and creating new uses is critical for an organization to survive into the future. Improvement of old systems rather than development of new systems may be the norm. Economics dictates making do with what already is available. This task is accomplished by targeting innovative enhancements of existing systems as a major method to satisfy requirements.

Organizations having flexible structures with as few layers as possible are best able to rapidly respond to the customer's changing demands. Organizations with rigid structures cannot react fast enough to keep pace with a formidable competitor. Few organizational layers provide the "lean and mean" structure needed to contend in tomorrow's world. Customers will no longer subsidize the cost of the huge waste created by large bureaucratic organizations. Organizations must be trimmed to the absolute core with decentralization of empowerment to the people closest to the customer and the processes. The organizational structure of an achieving organization has fewer managers with increased self-management.

Cooperation among governments, industries, companies, organizations, teams, and individuals is essential for survival. Cooperation among government, industry, labor, and education is critical to a high-growth, high-wage economy. In addition, management and labor must learn to cooperate for a prosperous economy. Further, departments and functional organizations must break down barriers to optimize productivity of an organization. Also, individual people need to work together in teams to rapidly respond to customers. In addition, organizations must develop supplier partnerships and customer relationships. All cooperative efforts aim at win-win solutions instead of a win-lose situation as fostered by competition. Only through cooperative relationships can global success be realized.

Groups, especially teams, are the organizational structures of choice. Although individuality is important, teams multiply the capabilities of each of the individual team members. In today's complex workplace, teams are the only structure capable of providing the high level of performance, flexibility, and adaptability necessary to rapidly respond to customers and provide deliverables that delight them.

People are the most important, flexible, and versatile resource capable of adding value to a product or service. Empowered people are the only resource with the ability to respond quickly to a customer by optimizing the output of a process based on a thorough analysis of customer requirements and the process. Therefore, specializing or eliminating people greatly reduces an organization's ability to keep or gain customers, which significantly decreases its chances for survival. People are the most important resource to gain an advantage over competition. To optimize this essential resource, forward-thinking organizations must strive to provide a high-quality work environment, where both the people's and organizations' needs are satisfied while they are striving to delight customers.

Strong leadership, both at the top and all other levels, is needed instead of strong management. Guiding people to achieve a common goal is the focus of improved performance in any organization. Strong management is still required to ensure that the project is completed as required, but leadership is essential to maximize the human potential to care about and satisfy customers. Managers simply ordering accomplishment will not make an excellent organization. Leadership involves the sustained, active, hands-on participation of all leaders continuously setting the example, coaching, training, and facilitating empowered people.

A long-term view needs to replace the emphasis on short-term results. Frequently, the emphasis on a short-term outcome has a long-term consequence. For example, the takeover of one organization by another may bring short-term financial rewards. In the long-term, the takeover can result in many people losing their jobs. In another case, the short-term focus on stocks causes the organization to reduce investment in capital equipment and/or training of its people. This could have an adverse effect on the long-term survival of the business. The viewpoint of the organization must be targeted on the long-term future for them to stay in business. The advanced organization has a vision for the future with a strategic plan for achieving it.

The reward and recognition system must shift from individual merit to team performance systems. A team performance reward and recognition system provides the incentive for optimizing the results of teamwork to accomplish the mission. The team performance system should credit the individual's contributions to the success of the team while at the same time providing a reward for effective teamwork. All reward and recognition systems should be geared to each individual team member wanting to contribute to the best of her or his abilities for the ultimate successful outcome of the team.

An education and training investment for everyone in the organization must be the cornerstone of any organization using a continuous improvement system to persevere in today's environment. People must be continuously improved through education and training with an eye toward the future. High-growth labor markets demand people with specific higher-level education and skills. These conditions require organizations to adopt a viewpoint that people truly make the difference in the competitive world economic environment, and they must make an investment to create a lifelong learning system.

Customer satisfaction through its deliverables and processes must drive the organization. The focus on profit as the primary purpose of an organization is obsolete. As shown in Fig. 1.2, the successful organization strives to meet customer expectations through the continuous improvement of process, people, and product focusing on total customer satisfaction. This effort promotes the use of best business practices

Figure 1.2 Chain reaction for success.

which leads to excellent business systems—which provides total customer satisfaction. The chain reaction starts with customer expectations, with total customer satisfaction being the focus of all efforts. This setup results in a successful organization. The definition of success, which varies by organization, is the vision. It can mean survival or growth, profits, return on investment, or stockholder dividends. The view of total customer satisfaction applies to both external and internal customers. The entire organization must be aimed at satisfying the ultimate user of the product or services. Within the organization, each process pleases the next process in the chain. Everyone works with an internal supplier and customer relationship to improve their process for total customer satisfaction of both internal and external customers.

Focus on the future

> *TQM establishes the flexibility necessary for long-term success.*

Total Quality Management focuses an organization on the future. Although every organization is unique, there are certain objectives,

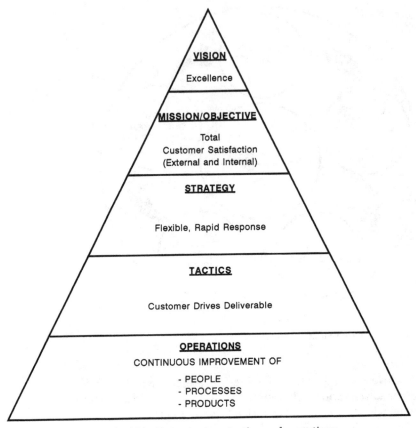

Figure 1.3 Vision, mission/objectives, strategy, tactics, and operations.

strategies, tactics, and operations that will work for organizations striving for a vision of the future. Figure 1.3 shows the vision, overall mission and/or objective, strategy, tactics, and operations necessary for any organization to prosper today with a view toward the future.

The vision is always striving for excellence. The objective is total customer satisfaction, both internal and external. To achieve this goal, customer expectations must be met or exceeded. This effort requires establishing and maintaining a Total Quality Management organizational culture focused on doing whatever it takes to add value for the customer. To create a Total Quality Management organizational culture requires vision and the leadership to make the vision a reality; involvement of everyone and everything in a systems approach; continuous improvement of people, processes, and products; training and education of everyone; ownership with empowerment; an appropriate rewards and recognition system; and years of commitment and support—all aimed at total customer satisfaction.

The strategy is flexible, rapid response, which requires an organizational structure featuring leadership and empowered people. The organization and its people must completely understand the customers and their processes. The organization must enable development of partnerships with suppliers and a relationship with customers which allows the organization to adapt their processes immediately to any of the customers' desires.

The tactics include the customer or customer's voice driving the deliverable (product and/or service). The process involves the organization providing the best possible deliverable at an optimal life-cycle cost. The deliverable, whether a product, service, or both, must be world class to compete in the global economy.

The operations include continuous improvement of processes, people, and products. Continuous improvement involves the elimination of loss in all processes, investment in people resource, and constant upgrade of products.

TQM gets results

TQM means satisfied customers at competitive cost.

Total Quality Management has transformed some of America's competitor nations into economic powers. In addition, many American institutions have already used TQM to pursue victory. In industry, many organizations have used TQM to increase customer satisfaction.

There are numerous examples of the success of TQM. Many seminars, symposia, books, magazines, newspapers, newsletters, and other media provide a never-ending flow of success stories. Most organizations freely share information about TQM and advocate its virtues. These stories prove that TQM works in both the private and public sectors. The Malcolm Baldrige National Quality Award winners provide recent excellent examples of success stories in America.

In general according to GAO/NSIAD-91-190, organizations that use TQM have found the following:

Greater customer satisfaction

Improved product and/or service quality and lower costs

Better employee relations

Increased financial performance

TQM Results

Raised productivity and quality

Enhanced ownership and empowerment of workers

Success for the organization

Useful systems, processes, and operations

Less cost and waste

Total customer satisfaction

Solved problems

Total Quality Management Definition

> *Doing the RIGHT thing,*
> *RIGHT the FIRST time,*
> *Doing it on TIME,*
> *ALL the time,*
> *Always striving for IMPROVEMENT,*
> *Always SATISFYING the customer.*

A simplified definition of Total Quality Management is: A leadership and management philosophy and guiding principles stressing continuous improvement through people involvement and quantitative methods focusing on total customer satisfaction.

This definition is the basis of this guide. This is the author's applied definition of TQM. There is no one universal definition of TQM accepted by everyone. The definition, as applied, of TQM can vary by organization and individual. This applied definition of TQM is usually internalized within each specific organization and every individual in the organization. Although the applied definition varies, every definition of TQM must include all the essential elements of TQM. The essential elements of TQM are

- Continuous improvement
- People involvement
- Use of quantitative methods
- Customer focus

Total Quality Management Philosophy

> *The TQM philosophy encompasses new management thinking.*

The Total Quality Management philosophy provides the overall, general concepts for a continuously improving organization. The TQM philosophy stresses a systematic, integrated, consistent, organization-wide perspective involving everyone and everything. There is much waste in organizations because of suboptization.

TQM focuses primary emphasis on total customer satisfaction (both internal and external customers). The key is to listen to customers, then satisfy their needs and expectations. Satisfied customers return and tell others.

TQM stresses a management environment fostering continuous improvement of all systems and processes. This initiative requires the institution of a continuous improvement system. Although this effort requires an investment in both time and resources, it ultimately leads to higher revenues and less cost.

The TQM philosophy emphasizes the use of all people, usually in teams, predominately with a multifunctional emphasis, to lead improvement from within the organization. The organization must get its own organization in order first, but it must always focus on customers. Improvements target productivity, quality, and cost management.

TQM stresses optimum life-cycle cost. Once this objective is achieved, savings over the long term can be made in failures, warranty claims, complaints, liability costs, and possible lost customers.

The TQM philosophy uses measurement within the disciplined methodology to target improvements. By measuring the right things, the organization pays attention to critical factors and the important actions get done.

Prevention of defects and quality in design are key elements of the philosophy. Elimination of losses and reduction of variability are its aims. These achievements save the organization on test and inspection, rework, scrap, repair costs.

Further, the philosophy advocates developing relationships—internal, supplier, and customer.

Finally, the philosophy includes an intense desire to achieve victory. Without this one element the others elements of the philosophy are limited. Typically, an organization needs years to make significant impact on the 20 to 60 percent or more of waste in most organizations. The organization needs the long-term commitment and support, which can only be fostered through a burning desire for results, to sustain the effort long enough (two to six years) to make TQM a way of organizational life.

In summary, the TQM philosophy includes

Pursue an organizationwide perspective (systems view).

Have a customer focus.

Institute continuous improvement of processes, products and/or services, people.

Lead improvements from within.

Orient everyone to drive improvements.

Stress optimum life-cycle cost.

Observe measurements and/or metrics.

Prevent defects, errors, waste, and design in product quality.

Have suppliers involved in improvement process.

Yearn for success.

TQM Guiding Principles

> *Principles or values indicate the basic beliefs that the organization follows.*

The TQM guiding principles are the essential, fundamental rules required to achieve victories. The TQM guiding principles involve continuously performing the following actions:

Provide a TQM environment.

Reward and recognize appropriate actions.

Involve everyone and everything.

Nurture supplier partnerships and customer relationships.

Create and maintain a continuous improvement system.

Include quality as an element of design.

Provide training and education constantly.

Lead long-term improvement efforts geared toward prevention.

Encourage cooperation and teamwork.

Satisfy the customer (both internal and external customers).

TQM stresses continuous improvement

> *Good, better, best, never let it rest; until the good becomes better and the better best.*

Continuous improvement is the only way to remain competitive over the long term. Profit not only requires revenue from customers but it also mandates cost management (profit = revenue minus cost). Continuous improvement focuses the organization on satisfying customers (generating revenue by getting and keeping customers) through the performance of its internal operations in a cost-effective way (optimizing costs through competitive internal processes, product and/or services, and people resources).

Continuous improvement of people, products and/or services, and processes is one of the essential elements of TQM. Continuous improvement involves using a disciplined approach to constantly make the organization excellent. The target of continuous improvement is quality, productivity, and cost management results.

Continuous improvement considerations

Improve, invent, reengineer, as appropriate

Make productivity and quality an obsession

Promote a disciplined approach

Recognize the value of people

Orient toward the customer

Visualize the ideal, long-term, total organization

Encourage relationships between owners, suppliers, and customers

TQM advocates people involvement

> *People make it happen.*

People are a competitive advantage for today's organization. The organization that maximizes its human potential will be a winner. Achievement of this objective requires the organization to encourage active participation of all the people in the organization. People involvement encompasses all activities to empower the people in the organization to achieve organizational results, including individual and team efforts.

People involvement considerations

Pursue a positive work environment

Encourage participation by everyone

Open communication

Provide an organizational system that meets individual needs

Lead by example toward a common vision

Enhance trust, cooperation, and teamwork

TQM uses quantitative methods

> *Measurement provides visibility for right actions.*

Quantitative methods allow the organization to know how it is doing and it gives the organization the facts for decision making. In most organizations, measurement directs action. Quantitative methods focus attention on the critical factors for success. It is important to measure to drive appropriate actions.

Measurement considerations

Meaningful for expected outcomes

Encourage constant improvement

Attention drawn to critical factors

Set by the people who are closest to making it happen

Use only as appropriate for organizational needs

Recognize the limits

Emphasize consistency

TQM has customer focus

> *Customers allow an organization to exist.*

The purpose of an organization is to satisfy customers. An organization depends on customers. Satisfied customers generate revenue or need for the organization. So any organization must have customers to survive. An organization cannot exist without customers. The bottom line in any organization is that it must satisfy customers in a cost-effective way to stay in operation. This fact is true for any organization large or small, private or public, profit or nonprofit, department or company.

Customer considerations

Can make or break an organization

Unique needs and expectations

Set requirements for quality as defined by their satisfaction

Treat customers as special

Orient everyone in the organization with a customer focus

Make getting and keeping customers the target

Encourage the "voice" of the customer in product and/or service design

Relate to customers as long-term partners

The TQM Umbrella

> *TQM considers all tools and techniques that can help the organization improve.*

The TQM umbrella includes the integration of all the fundamental management techniques, existing improvement efforts, and technical tools under a disciplined approach focused on continuous improvement. All existing improvement efforts fall under the TQM umbrella. Figure 1.4 shows some of the current best known improvement efforts. These efforts include concurrent engineering, robust design, statistical process control, just-in-time, cost of quality, total production maintenance, manufacturing resource planning, computer-aided design, computer-aided engineering, computer-aided manufacturing, computer-integrated manufacturing, information systems, total integrated logistics, and total customer service.

Seven Basic Quality Tools

Manufacturing Resource Planning

Just-In-Time

Total Production Maintenance

Statistical Process Control

Seven Management Tools

Information Systems

Total Customer Service

Total Integrated Logistics

Robust Design

Computer-Aided Design, Engineering, and Manufacturing

Reengineering

Computer Integrated Manufacturing

Cost of Quality

Mistake Proofing

Project Management

Quality Function Deployment

Design of Experiments

Concurrent Engineering

Other proven tools and techniques

Figure 1.4   TQM umbrella.

The improvement efforts target improvement of one or more of the aspects of an organization. For instance, robust design, statistical process control, just-in-time, cost of quality, total production maintenance, manufacturing resource planning, computer-aided design, computer-aided engineering, computer-aided manufacturing, and computer-integrated manufacturing are engineering and manufacturing oriented. These existing improvement efforts can show some visible results by themselves. However, TQM integrates all of these improvement efforts to enhancing the overall effectiveness of the entire organization focusing on customer satisfaction.

TQM, a unique management approach

TQM requires a break from approaches that no longer work.

TABLE 1.2 Comparison of Traditional Management and TQM

Traditional management	Total quality management
Looks for the "quick" fix	Adopts a long-term, strategically oriented philosophy
Firefights	Uses a disciplined methodology of continuous improvement
Operates same old way	Advocates breakthrough, innovation, and creative thinking
Adopts improvements randomly	Systematically selects improvement
Inspects for defects or errors for product quality	Focuses on prevention
Decides using opinions	Decides by using facts
Throws money and technology at tasks	Maximizes people resources
Controls resources by function	Optimizes resources across the whole organization
Controls people	Empowers people
Targets individual performance to meet job description requirements	Focuses on team performance to meet customer expectations
Is primarily motivated by profit	Strives for total customer satisfaction
Relies on programs	Is a never-ending process

TQM is a unique management approach. As explained in the philosophy and guiding principles, it is a change from the traditional American management mindset. To get a clearer understanding of what TQM is, Table 1.2 provides some comparisons between traditional management and Total Quality Management.

TQM is a people-oriented, measurement-driven, customer-focused, long-term, strategically oriented management philosophy using a structured, disciplined continuous improvement operating methodology. It is not a "quick fix" using fire-fighting techniques. TQM uses many small, continuous improvements targeting breakthroughs, innovations, and creativity rather than simply operating the same old way with business as usual.

With TQM, management must systematically select long-term continuous improvement efforts. In the past, many improvement efforts were randomly adopted in the organization for a short period.

TQM focuses on "doing the right thing right the first time." This focus emphasizes prevention of errors and quality of the design. Traditionally, inspection was the method used to find and eliminate errors.

Further, TQM bases decisions on facts rather than opinions as traditional management often does.

TQM's use of people's capabilities as a primary means to add value to the product or service is also a major variation from the traditional approach. Normally, management tends to increase resources or technology to add value to its product or service.

With TQM the objective is to optimize resources across the whole organization. Traditional management suboptimizes resources by function.

TQM fosters the empowerment of people to perform and improve their processes rather than controlling people.

Team performance focused on total customer satisfaction is valued, rewarded, and recognized in TQM. Traditionally, individual performance targeted to meeting specific job descriptions was the organization's goal.

Above all, the TQM philosophy focuses on customer satisfaction. It is not solely motivated by profit.

Finally, TQM is not simply a new management program; it is a never-ending way of life for the future.

The Total Quality Management Process

> *The TQM process starts and ends with the customer.*

Total Quality Management focuses on the continuous improvement of all systems and processes in the organization. In fact, TQM is a process itself. TQM is a process within the overall system of the organization. The entire organization is a system made up of many processes to accomplish the functions of the organization.

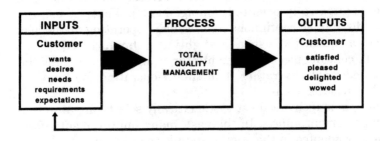

The TQM process starts and ends with the customer.

Figure 1.5 TQM process.

But what is a process? A process is a series of activities that takes an input, modifies the input, and produces an output. The TQM process transforms all the inputs in the organization into a product and/or service that satisfies the customer. In Fig. 1.5, the overall TQM process consists of the inputs received from a supplier, the process itself, and the outputs supplied to the customer. The most important inputs include the wants, desires, needs, expectations and requirements of the customer. These inputs are combined with many other inputs to the process, including people, material, supplies, methods, machines, technology, and the external environment to create an internal TQM organizational culture and a deliverable. The output of the process is increased financial performance, improved operating procedures, better employee relations, and greater customer satisfaction. The TQM process starts with the customer and its output focuses on the customer.

2

The TQM Foundation

TQM is built on a foundation of ethics, integrity, and trust with open and honest communication. It requires an organization to change. It must have a good business reason for doing it. Finally, the organization must be ready for TQM.

Before embarking on the TQM journey, the organization needs to answer the following questions:

- Does the organization have a foundation for TQM?
- Can the organization deal with the change necessary for TQM?
- Is TQM the right approach for the organization?
- Is the organization ready for TQM?

TQM Foundation Considerations

Foster openness, fairness, and sincerity.

Operate with honesty.

Use common sense.

Nurture trust.

Demonstrate appropriate behavior.

Allow involvement by everyone.

Teach right from wrong.

Instill values into organization.

Only do to others what you would want done to you.

Never compromise ethics, integrity, or trust.

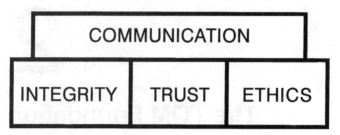

Figure 2.1   The TQM foundation.

TQM Foundation

An organization is only as strong as its foundation.

To enter into Total Quality Management requires a foundation of ethics, integrity, and trust with open and honest communication. This is the key to unlocking the ultimate potential of TQM. An organization's potential for high performance in both good and bad times improves with a solid foundation of ethics, integrity, and trust. Figure 2.1 shows ethics, integrity, and trust as the foundation, with trust binding ethics and integrity. Communications is the link of the foundation into many other parts of the structure.

Each organization contemplating TQM should take the time to lay the foundation. Since everything builds on the foundation, ultimate success or failure depends on the foundation. Many organizations have forgotten this simple fact.

Any discussion of TQM that does not include a presentation of integrity, ethics, and trust would be greatly remiss; in fact, it would be incomplete. TQM is built on a foundation of integrity, ethics, and trust. If VICTORY is to be achieved, these characteristics must be a vital part of TQM. Total Quality Management requires the active participation of everyone in the organization focusing on customer satisfaction. Business is maintained and won through customer satisfaction; it is not achieved by bad business practices or profit at any cost. This aim demands that everyone in the organization have the highest standards of integrity, ethics, and trust in dealing with both internal and external customers.

Integrity, ethics, and trust move together through the TQM environment. However, each element offers something different to the TQM concept.

Ethics

> *Do to others only what you would want done to you.*

Ethics is the discipline concerned with good and bad in any situation. Ethics is a two-faceted subject represented by organizational and individual ethics. In the case of organizational ethics, most organizations will establish a business code of ethics that outlines ethical guidelines that all employees are to adhere to in the performance of their work. Individual ethics includes personal rights or wrongs. They are concerned with legal, moral, and contractual business policies and individual dealings. A person should never do anything that goes against command media or that the person would not like done to him- or herself.

Integrity

> *A person's true nature will reveal itself despite disguise.*

Integrity implies honesty, morals, values, fairness, adherence to the facts, and sincerity. This characteristic is what anyone in the organization and the customer (internal/external) expects and deserves to receive.

True nature cannot be disguised. People see the opposite of integrity as duplicity. TQM will not work in an atmosphere of duplicity. In a deceptive environment, the organization takes actions contrary to the philosophy and guiding principles of TQM. If anyone in the organization perceives contrariness of thought or action, active involvement will not be accomplished. In addition, if the customer perceives the organization as being guilty of duplicity, customer satisfaction will not be achieved and quite likely the customer's business will be lost.

Trust

> *Trust others and they will trust you.*

Trust is a by-product of integrity and ethical conduct. It is absolutely essential for TQM. Without trust, the framework of TQM cannot be built. This quality is important in all aspects of TQM, including teamwork, improvement efforts, and customer satisfaction. Trust starts with open and honest communication, which is necessary for involvement of the right people in the TQM process. Trust fosters full participation of all members. It allows empowerment that encourages pride of ownership, and it encourages commitment.

Specifically for TQM, trust promotes teamwork and cooperation. It allows decision making at appropriate levels in the organization; fosters individual risk taking for continuous improvement; and helps to ensure that measurements focus on improvement of process and are not used to control people. Trust is essential to ensuring customer satisfaction.

For the traditional conflicts in an organization to be removed, trust must be developed. This quality is necessary to change the adversarial relationships between company and union, management and labor, and different functions in the organization. Trust builds the cooperative environment essential for TQM.

Communication—the vital link

> *Actions speak louder than words— Talk the talk and walk the talk.*

Communication is the most important tool in any organization. In a TQM organization, communication is essential. The scope and level of communication increases in a TQM organization. Communication involves the exchange of information, and TQM demands a free flow of information. The success of TQM demands communication with and among all organization members, suppliers, and customers. It also requires frequent and effective communication with people. It demands communication both inside and outside the team. The organization needs information to understand the needs and expectations of the customer. Organization members need information from each other to complete and improve processes. They rely on information from support teams. There must be constant communication between customers, process owners, program managers, suppliers, other support teams, and the functional organization. Communication coupled with the sharing of the right information is vital.

Communication of the right information is a complex process which involves many verbal and nonverbal forms including speaking, listening, observing, writing, and reading. Because of this complexity, the information may not be communicated correctly. Even in the simplest communication model with just a sender, message, and receiver, there are many obstacles to effective communication. For communication to be effective, the sender must be credible, the message must be clear, and the receiver must interpret it the way the sender intended. For example, if the sender is not trusted by the receiver, the sender may not be able to communicate with the receiver. Regardless of the message, communication will be ineffective.

Communication gets even more complex if we add reality to the model. Rarely do we communicate with just a sender, message, and receiver. Normally, there are many distractions. We may be influenced by our work environment (e.g., political pressure, fear) or thinking about other things at home or at the workplace. We also have different values, cultures, perceptions, and so on. Communication can be improved by doing the following:

Clarify the message.

Observe body language.

Maintain everyone's self-esteem.

Make your point short and simple.

Understand others' points of view.

Nurture others' feelings.

Involve yourself in the message.

Comprehend the message.

Attend to the message of others.

Talk judiciously.

Emphasize listening.

Feedback

Because of the possibility of ineffective communication, it is critical to ensure through feedback that the right information is communicated. It is always the responsibility of the sender to ensure effective communication. Feedback involves providing information back to the sender to verify the communication. Feedback can indicate agreement, disagreement, or indifference. Like communication, it can be verbal and/or nonverbal. Some guidelines on effective feedback follow:

Foster an environment conducive to sharing feedback.

Encourage feedback as a matter of routine.

Establish guidelines for providing feedback.

Discuss all unclear communications—paraphrase and summarize.

Be direct with feedback.

Ask questions to get better understanding.

Consider "real" feelings of team members.

Keep focused on mission.

Listening

Listening is a technique for receiving and understanding information. It is one our most important communication needs, but it is the least developed skill—although critical to effective teamwork. Effective listening requires an effort to understand the ideas and feelings the other person is trying to communicate. An effective listener hears the content and the emotion behind the message, attending not only to what the person is saying but also to gestures, posture, and vocal qualities. Expert listening requires active behavior, making an effort, and paying attention to the person and the message. It means actively communicating that you are listening and trying to understand the other person. Discipline, concentration, and practice are essential. Effective listening requires the following:

Let others convey their message.

Involve yourself in the message.

Summarize and paraphrase frequently.

Talk only to clarify.

Empathize with others' views.

Nurture active listening skills.

When listening, let the other person convey her message without interrupting or forcing your own views. You can do this by letting the other person know you are interested in what she is communicating without displaying an opinion or judgment.

Involve yourself in the message by actively listening to what the other person is communicating. Establish and maintain eye contact. Keep an alert posture. Look for verbal and nonverbal cues.

Summarize and paraphrase frequently to show an understanding of the message. By listening carefully and then rephrasing in your own words the content and feelings of the other person's message, the exact meaning of the message can be determined.

Ask questions to clarify points you do not understand. Points can be clarified by using open-ended questions. This type of question which requires an answer other than yes or no provides a more detailed explanation.

Understanding the other person's views is essential to effective listening. Set aside your opinions and judgments and place yourself in the other person's place. Show him you understand by requesting more information or by sharing a similar feeling or experience you've had and explaining how you think it helps you understand him.

Nurture listening skills to improve communication. Listening skills must be practiced daily.

The TQM Foundation Supports Change

> Change is a constant.

A foundation of ethics, integrity, and trust provides a safe environment for change. The presence of a solid foundation is critical during change. The foundation not only supports but also protects. It supports the business's systems, including TQM. At the same time, the foundation protects the organization from an adverse environment.

In every organization change is necessary to establish the TQM foundation and implement TQM. The nature of TQM requires both the individual in the organization and the entire organization to change. As seen in Chapter 1, today's and tomorrow's world demands change.

Chapter 1 outlines some of the changes your organization may be experiencing. In addition, individuals in organizations are experiencing many changes which may include

- Different roles
- Working in groups or on teams
- Empowering others
- Self-inspection or self-autonomy
- Sharing decision making
- Delegating or taking more responsibility
- Learning new skills
- Constant pressure to perform and improve processes
- Leading toward a common purpose
- Becoming more customer oriented

- Developing creativity
- Building flexibility and adaptability
- Balancing many priorities
- Managing conflict
- Maintaining relationships

Dealing with change

As everyone is acutely aware today, change is part of the everyday environment in any organization. Total Quality Management increases the intensity of change in the organization. Therefore, it is important for every organization, especially an organization embarking on the TQM journey, to learn to deal with change.

TQM change must start with the leadership in the organization and then target achieving a critical mass to be successful. Change should never be allowed to just happen. Simply announcing a change will not make it happen. Guiding change using proven leadership and management techniques is a necessity for desired outcomes to be achieved. First, change must be led through specific direction and communication. Second, coaching instills the desired change. Third, the change must be supported through a formal system. Finally, the change is part of the organization that continues to be monitored for possible continuous improvement.

Change must always be viewed within a time reference. Today, more than ever, time is always a deciding factor. The speed of the change depends on the situation and the organization. Some organizations can afford to take their time and they may use an evolutionary approach to change. Other organizations' survival depends on rapid change. These organizations may use a more radical approach.

In addition, an organization's change performance depends on its ability and willingness for a particular change. The ability to make a change depends on the resources of the organization. The organization must make the necessary investment to give the change a chance. The willingness of the organization is related to the sense of urgency or common purpose. The more the people in the organization relate to the need for the change, the better the potential for the desired change.

Change, especially with TQM, must focus on improving processes. In TQM, the change process targets demonstrating to people in the organization that TQM does work and it is better for them and the organization than the old process.

Dealing with change requires leadership in the organization to do the following:

Communicate.

Have structured activities.

Acknowledge people concerns.

Nurture individual differences.

Get focused on real issues by setting priorities.

Encourage creativity and innovation.

Set a positive example.

Communicate the focus for the change. Communication is the key to success in most human endeavors. Communication allows people to understand and deal with the change.

During change people often feel stress, insecurity, awkwardness, embarrassment, frustration, confusion, anger, joy, happiness, and so on. It is critical for each person to understand how the change affects him or her personally. In the early stages of change, people are not interested in the benefits to the organization. It is critical to communicate early with the people undergoing change to let them discover what's in it for them. Communication helps each person understand the impact of the change on her or him personally.

When TQM is being implemented, each person in the organization must determine the effect of the change in his or her own terms since various people are motivated differently. Some people look at the positive aspects of change. They see the opportunities for growth and the challenges of striving for excellence. Others only see the possible negative consequences; they might be roused by the possibility of not losing their jobs as a result of TQM.

As the change or TQM effort continues, it is essential to communicate by every verbal and written means possible specifically what is happening and what is expected of people in the organization. In all communications, try to be at ground level as much as possible, especially in the early stages. Lofty or vague communication only makes things worse. For instance, statements such as "you are empowered," "be a team player," and "focus on the customer" are not understood by people in the organization. Therefore, instead of rallying the people to action, these statements usually freeze people into little or no action. The organization must communicate the operational definition in specific, understandable terms to produce results and overcome resistance. Once the organization is comfortable with change, communications can evolve to a higher level to create a challenge and foster innovation.

Have structured activities. Even when everyone in the organization is undergoing change, people frequently feel alone. Structured activities not only allow people to discover they have a lot of concerns in common but they also provide a forum for positively dealing with the change.

In a TQM effort, it is essential to use an organizationwide, systematic, integrated structured approach. This process allows people in the organization to see how each piece contributes to the success of the whole. It also helps create the synergy necessary to focus everyone together for one common purpose. This effort prevents results from being isolated into islands of failure or success.

In the beginning of any change effort, including TQM, it is absolutely necessary to bring all the stakeholders together to get support and commitment. This support leads to ownership for the change. Once stakeholders accept ownership for the change, they will do whatever is necessary to ensure success.

During the change, ensure that support groups will be available to help various people in the organization deal with the change. For instance in a TQM implementation, everyone in the organization is affected by change in a different way. It would be appropriate to have structured activities for executives, middle managers, supervisors, and workers to help each group deal with changes. In TQM organizations, a formal support structure may need to be in place to facilitate the TQM process. In addition, TQM often requires systematic education, training, facilitating, coaching, and mentoring activities. Further, it is helpful to structure activities into projects.

Acknowledge people concerns. People have various concerns in any change effort. They are worried about jobs, finances, routine, relationships, and so on. These personal concerns must be dealt with immediately and directly. It is imperative to deal with people's concerns and provide as much support and assistance as possible.

Although you must deal with people's legitimate concerns, do not become distracted by resistance tactics. Since most people know that many change efforts go away after a short period of time, usually less than six months, people who resist will use many tactics to stop or delay the change. Some of the many resistance tactics can be described as follows:

- Endless discussion
- It won't work here
- Same old way
- Not invented here
- Shoot the messager

- I'm too busy, or I've got too many important things to do
- Marching in place
- Endless discussion
- Analysis paralysis
- Agreeing but not committing
- Finger pointing
- Emotional displays
- Saying one thing and doing another
- I don't understand
- Just another program or fad
- TQM tool of the month

Resistance must be met with a firm reaffirmation of the change in a pleasant and positive manner. Repeated resistance to the change should first be met with an ultimatum. If the person doesn't do some positive action, then the person will have a consequence. If the person then doesn't meet the request, appropriate actions must be taken against that person.

A TQM effort is no different from any other change; it requires support of people's legitimate concerns, but it also should have a positive straightforward focus on doing what is necessary for the overall good of the organization.

Nurture individual differences. Each person and organization has their own capacity for change. Therefore, each person reacts differently to change. Some people welcome it. They need change to be excited, enthusiastic, and motivated and to feel a sense of accomplishment. Others are overwhelmed by even a minor change in their routine. Still other can accept some changes up to a certain limit.

Individual differences and organizational capacity for change require management of change at the optimum organization and individual levels during any TQM effort. The organization as a whole cannot allow some people to move ahead while others are slow or immobile. These reactions to change have an adverse affect on the change effort. Ideally, the organization needs to move slowly and steadily to win the race. Some of the more successful organizations with TQM take from 6 to 18 months to prepare for the implementation of TQM by developing an understanding of TQM, building the foundation, and creating the framework. However, once TQM is initiated into the organization it is important that some positive, visible, practical small wins are achieved early in the process. In all organizations, the pace of the TQM efforts should be geared to the urgency for the change and the organization's capacity to make the necessary changes.

Individual differences for change can be optimized to support the overall change effort. People in the organization with the ability and willingness to deal with a large amount or rapid speed of change should be sought to take on the champion roles in the TQM effort. They can be used to coach, train, facilitate, or mentor others in the organization, allowing them to make one more change effectively. This process slows the overall organization rate for change while increasing the limits of change for others in the organization. Thus, a steady pace for the organizational change is maintained.

Get focused on real issues by setting priorities. During a TQM effort, people can become disoriented and overwhelmed because sometimes the emphasis shifts to the changes necessary for TQM while at the same time the daily work continues. Leadership must establish a common focus to help everyone in the organization know what is important.

It is critical to TQM that people integrate TQM into their everyday work life and organizational culture. TQM must become the way of life in the organization. The organization is then allowed to focus on critical issues and set priorities.

Encourage creativity and innovation. In every organization, people are already doing the best they can with the resources available. During change, both the organization and the individual are expected to produce more goods or services more frequently with the same or fewer resources. If they continue to do things the same old way, they can only get the same output. Therefore, the organization must encourage creativity and innovation to find new or different ways to do things.

In a TQM effort, creativity and innovation are especially critical. Traditional approaches no longer work. People need to do things differently. They need new systems, processes, methods, tools, and techniques. Most importantly, the organization needs to mine the diversity of the work force for ideas.

As part of encouraging creativity and innovation, the organization must promote risk taking. Without a tolerance for failure, people will not take the necessary risk to be creative or innovative.

Set a positive example. It is essential during any change or TQM effort to set a positive example. During any change effort once people sense any wavering of focus or commitment, they try to return to what was comfortable for them. Depending on the length of time the change effort has been operational, people will react in a number of ways: at best, they will go back to doing exactly what they did before the change effort was started; at worst, they will have forgotten how to do what they did exactly as before, and the result will be output *lower* than before the change effort. Therefore, it is critical once a change or TQM effort is started for the pressure to stay on until people learn the new way of doing it.

Is TQM Right for Your Organization?

> *TQM must be justified by good business sense.*

The question above is a critical one to answer. Although TQM can be adapted to any specific organizational situation to improve that organization, it may or may not be right for *your* organization at this time for a variety of reasons. You should not pursue TQM unless there is some potential for success.

TQM is definitely not appropriate for your organization if

- It is too late in a crisis situation or you are too far behind to try to do anything.
- There is no business or organization reason for using TQM.

In the situations above, it is clear that the organization should never consider TQM because it would not be successful. For example, if the organization is already doomed, TQM will not awaken the dead. Also, if the organization does not have a clear and valid justification, TQM will not be successful.

TQM may not be right for your organization at this time if

- The organization does not have the proper TQM foundation.
- The organization cannot deal with any more change.
- The organization is not ready for TQM.
- Top management does not support it.
- The organization does not have the ability and willingness to commit resources.
- The organization is just looking to the latest management fad to implement as a program.
- The organization expects TQM to be used for cost cutting only.
- People are not valued as a competitive resource.
- The organization only wants product or service quality.
- Your industry or organization is stable, predictable, noncompetitive.

In the cases above, the organization may decide to do some other organizational or individual interventions or development activities before embarking on a TQM journey. For TQM to be successful in these situations, the organization would have to do preparation activities to pave the way for TQM. For instance, an organization might learn to increase its capacity for change, thus making TQM feasible. If an organization facing one or more of the situations above embarks on TQM without considering the additional requirements, they appreciably decrease the probability for success. Each organization needs to make its own decision to pursue TQM or some other form of organization or individual development. The organizational or individual development could be accomplished prior to TQM. In some situations, an organization could decide to pursue TQM while at the same time considering use of other organizational development activities. For example, an organization could first pursue a pilot project to get a success with a change effort before it fully invested its resources in TQM. Another organization may choose to seek an ISO-9000 certification or Malcolm Baldrige Award while installing the TQM process.

Why use TQM for your organization?

> *Each organization must have its own reason for TQM.*

If, after careful evaluation, the organization decides TQM may be right for it, it must still as an organization specifically answer the question, Why use TQM? Generally, there can be a case made for TQM as the means for surviving and thriving in any organization. In fact, for most organizations today, use of TQM is a matter of continuing to exist. For a few others, TQM provides the means for thriving in the future. In either case, the reasons for using TQM need to be as specific as possible to gain the required support and commitment in the organization.

TQM should only be considered in an organization if it is the means to a specific end in that particular organization. It should not be pursued because "everyone is doing it" or "the boss wants quality." It is no longer just a "quality" thing. Product or service quality is imperative just to play the game in today's world. Competitive advantage is key. Pursuit of quality is an ongoing endeavor that goes beyond simple product or service quality to "quality" as defined by the customer. This effort requires organizations to compete on many levels such as cost, technology, quality, service time, and value—many of these at the same time. Although TQM can make an organization more competitive on any one of its many competitive realms, the essence of TQM requires an intense, common organizational focus based on each organization's specific determination of total customer satisfaction for success. All the systems and processes within the organization need to be geared toward achieving the specific results needed to satisfy customers. There are many general reasons for using TQM (see Chapter 1), but each organization needs to determine its own specific ends for TQM. There must be one focus for TQM to create the sense of urgency and purpose for the successful TQM implementation.

Are You Ready?

Once the organization has justification for TQM, it needs to decide the readiness of the organization for TQM. Management must analyze the organization to determine the specific TQM requirements. The checklist in Fig. 2.2 will help determine the readiness of the orga-

ITEM	YES	NO
Is improvement necessary in the organization?		
Is the organization willing to change to make improvements?		
Does top management understand that change is necessary for improvement?		
Is there a sense of urgency from top management for creating the required changes?		
Can the sense of urgency be understood by the entire organization?		
Does the organization have or is it willing to create the necessary foundation of ethics, honesty, and trust?		
Is there a business reason for TQM?		
Is the organization willing to create an organization that has a primary customer focus?		
Are there or can there be sufficient champions in the organization willing to install the TQM process? (leadership)		
Does top management have a vision of where the organization should be in the future? (vision)		
Is management willing to mobilize everyone and everything to make the necessary changes? (involvement)		
Can an appropriate continuous improvement system be established throughout the entire organization? (continuous improvement)		
Is management willing to invest the funds, time, and resources necessary for appropriate people development? (training, education, coaching, etc.)		
Is management willing to develop the trust necessary to allow people to take ownership for their work? (ownership)		
Is management willing to install appropriate recognition and rewards system to foster expected behavior and results? (reward and recognition)		
Is management willing to commit years of support, funding, and resources for TQM? (yearning for success)		

Figure 2.2   TQM readiness checklist.

nization to begin TQM implementation. In addition, this checksheet provides an insight into what is required to implement TQM in a specific organization. This knowledge will give the impetus to do something to start the organization toward victory.

Building the TQM Foundation Action Process

The action process for building the TQM foundation is as follows:

1. Assess the current TQM foundation, including the communication link.

2. Evaluate the organization's capacity to deal with change.

3. Determine if it is the right approach for the organization.

4. Decide if the organization is ready for TQM.

5. Decide on the next steps.

Building the TQM foundation action process flow

The TQM foundation action process flow is illustrated in Fig. 2.3.

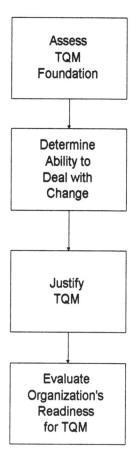

Figure 2.3 TQM foundation action process flow.

3

The TQM Framework

Customer focus + Leadership + Victory

Once an organization decides to pursue Total Quality Management, it needs to look at establishing the TQM framework. The TQM framework builds on the TQM foundation, providing an appropriate organizational environment for TQM. It starts with a customer focus, then builds with modern leadership that drives all the elements of VICTORY.

The TQM framework requires

Focusing on customer(s)

Instituting modern leadership

Visioning a common focus

Involving everyone and everything

Continuously improving people, processes, and product

Training, educating, coaching, facilitating, and mentoring

Owning the organization, TQM, processes, and work

Recognizing and rewarding

Yearning for success

TQM framework environment considerations

Empowering

Nurturing relationships

Visionary

Innovating

Rewarding, renewing

Opening opportunities

Nurturing pride of ownership

Meaningful

Enterprising

Nurturing openness and honesty

Trusting

TQM Framework

The organization must establish and maintain a TQM framework over the long term. This framework requires a systematic, integrated, consistent, organizationwide perspective. It won't just happen. The TQM framework must include the entire organization and be shared by everyone in the organization. Leadership is required to create a TQM framework focused on total customer satisfaction that fosters the attainment of VICTORY. This is the VICTORY-C model as shown in Fig. 3.1. All of the elements of VICTORY-C are absolutely essential for survival today and victories in the future. This TQM framework has the customer as the focus or center. It includes the application of leadership to achieve VICTORY.

For success to be achieved the elements of VICTORY must be manifested in every aspect of the organization. First, there must be a vision providing the purpose of the organization as viewed by top leadership and shared by everyone in the organization. In addition to a vision, there must be a common focus throughout the organization maintained and developed by the leadership. Equally essential is the involvement of everyone and everything and the continuous improvement of all systems and processes. A fourth necessity is the development of the people in the organization through training, education, coaching, facilitating, and mentoring. This discipline must be constantly provided for a learning organization. Fifth, ownership must be established and fostered for all systems and processes in the organization. Sixth, recognition and rewards must be systematized to rein-

Figure 3.1   VICTORY-C TQM model.

force desired behaviors and outcomes. And seventh, the organization, especially top leadership, must want success. VICTORY requires years of personal commitment and active support.

Customer focus

> *The primary purpose of any organization is to satisfy customers' needs and expectations.*

Customers allow an organization to exist. This fact is true of every organization, profit or nonprofit, company, partnership, sole proprietorship, department, function, group, or team. Therefore, a customer focus is one of the major elements of the framework of TQM.

All the elements of VICTORY focus on total customer satisfaction, external and internal. Total customer satisfaction is the focus of the entire TQM process. Its primary concern is with quality, including all elements required to satisfy the target customer(s), both internal and external. Examples of such elements are

- Product quality
- Service quality
- Performance
- Availability
- Durability
- Aesthetics
- Reliability
- Maintainability
- Logistics
- Supportability
- Customer service
- Training
- Delivery
- Billing
- Shipping
- Receiving
- Repairing
- Marketing
- Warranty
- Life-cycle cost
- Responsiveness
- Time
- Accuracy
- Dependability
- Consistency
- Ease in doing business with
- Doing what is promised
- Doing what is supposed to be done
- Providing value
- Low cost
- Superior technology

Figure 3.2 shows the relationships of internal and external customers. Each process is the customer of the next process. These are

Feedback

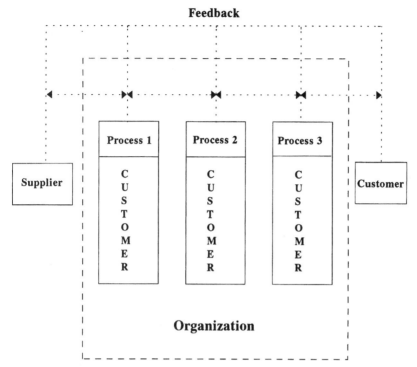

Figure 3.2   Customer relationships.

the internal customers. If each internal customer satisfies the next internal customer while focusing on external customer satisfaction, the ultimate customer—the external customer—will be satisfied. TQM teams assure that each process is linked to the customer.

Focusing on total customer satisfaction

Customers are the only ones who can determine total customer satisfaction. In order to know if the customer is satisfied, intense observation is necessary. Only through observation, communication (especially listening), and measurement can the organization determine total customer satisfaction. The organization must use every means available to evaluate customer satisfaction.

Customers are the focus of all Total Quality Management efforts. Therefore, every organization and everyone in the organization must constantly strive to satisfy the current customers and create new customers for the future.

Total customer satisfaction considerations

Supply product and/or service quality.

Allow flexibility and adaptability to customers' needs and expectations.

Think responsiveness.

Include value.

Strive for excellence.

Find "real" customers.

Achieve quality "image."

Communicate, especially listen, to customers.

Train for consistency in product and processes.

Include everyone in the organization.

Organize satisfying customers.

Nurture supplier and customer relationships.

Understand what it takes for total customer satisfaction

Besides an obsession for quality from the inside out, total customer satisfaction requires the organization to know itself, its product, its competition, and its customers. The first step to focusing on total customer satisfaction is to know your customers. This chapter describes the process of knowing customers, and Chapter 4 describes the process of knowing the organization itself, its products and competition.

Know your customers

Get to know your customers. It is not always obvious who your customers are. Customers are all those people touched by the product or service, internal and external to the organization. To continue to satisfy the customer, all customers must be identified.

Next, who the target customers are must be determined. Once the target customers are identified, customer needs and expectations must be determined. Customer expectations are dynamic; they continuously increase and change. So a continuous review of customer needs and expectations is necessary to ensure customer satisfaction. Like so much of TQM, this process, too, is continuous.

Understand customer needs and expectations

The identification of customer needs and expectations requires systematic, thorough, and continuous marketing research. The most important aspect of this process is to listen to the customer.

It is important to use as many tools and techniques as possible to understand your customers. Some of the most common marketing tools include media research, test marketing, customer auditing, and customer focus groups. In addition, each organization should consider TQM tools such as customer analysis and quality function deployment.

Changing customer needs

Customer needs are not static; they are always changing. Once customer needs are identified, they must be continuously monitored to ensure that the product and/or service still satisfies them. People have different needs, from basic survival needs, such as eating and sleeping, to total fulfillment of one's life goal. Customers satisfy lower-level needs before higher-level ones. A need once satisfied is often no longer a need.

Needs are constantly replaced by other needs due to the changing world environment. Rapidly changing technology, differing tastes, and rising expectations due to past successes are some of the many factors influencing customer changes.

Develop customer relationships

To ensure that the organization is continuing to satisfy the customer, a relationship with the customer is crucial. Relationships demand continuous attention. As Theodore Levitt states in *The Marketing Imagination,* "the sale merely consummates the courtship. Then the marriage begins. How good the marriage is depends on how well the relationship is managed by the seller. That determines whether there will be continued or expanded business or troubles and divorce, and whether costs or profits increase." The emphasis is on keeping current customers while seeking additional customers for the future.

Like all relationships, customer relationships require communication, support, and responsiveness. Communication, especially listening, is essential. The customer needs to be involved in as many aspects of the product as possible. Support must be available to help the customer with the product after it is received. Responsiveness is key to continuing the relationship. The organization must be able to respond to the needs of the customer in any situation.

Establishing a Customer Focus Action Process

1. Know your customers' needs and expectations.
2. Use information to develop a strategic plan that establishes a customer focus.
3. Develop internal process capabilities to meet customers' needs and expectations satisfactorily.
4. Monitor customer trends to make changes as appropriate.

Leadership

> *Lead and others will follow.*

Leadership along with customer focus form the two major elements for the TQM framework. Without these two elements, the elements of VICTORY will only show limited results. While customer focus provides the common focus or target for TQM, leadership is the driver. Without a destination and a driver, the organization will not go anywhere. It will stay where it is now.

Leadership is essential to make TQM a reality. The leaders of the organization set the example for actions throughout the organization, create the reason to work productively, inspire the imagination of the people in the organization, and bring meaning to the workplace by allowing each person to focus on his or her contribution to the greater societal good—the needs of the customer.

Role of modern leader

The modern leader faces many opportunities. In today's environment, the role of the leader is constantly changing to adapt to the varying external conditions and changing needs of the organization. This evolving role presents a particular challenge to many modern leaders. They must juggle the requirements of both strong management and leadership. Typically, management skills have been strong in most organizations. However, some organizations and managers are frustrated with their lack of leadership competencies.

In addition, the need for leaders has grown in recent times—and it is not limited to the upper levels of the organization. Many organizations require leaders of teams everywhere in the company. The team leader competencies vary according to the particular situation.

Because of the above factors, defining the modern leader is not an easy task. To complicate matters, although there are certain characteristics that make up a modern leader, there is not one set of competencies that make a leader successful in every organization. A leader's competency depends on the particular organization.

Leadership in a TQM framework is usually characterized by the following kinds of traits:

1. *Leaders set the example.* They live their vision by incorporating their visions into their daily behaviors. Leaders guide others by displaying behaviors and actions that direct people to their vision.

2. *Leaders create focus.* This focus is the vision of the future. Visions are pictures of how the world could be and allow people to connect their own contribution to the realization of the vision. The vision becomes meaningful to every member of the organization because they align themselves with it.

3. *Leaders create synergy.* This synergy is developed by the creation of meaning and value beyond the self and into the organization. Leaders call up the inherent needs of people to work cooperatively to create values, services, and products which go beyond the sum of the parts. Leaders inspire coordination, collaboration, and cooperation because they are successful in establishing a shared vision of progress and growth.

4. *Leaders establish and maintain organizational systems.* These systems must support the TQM framework. The organizational systems must promote the maximizing of people resources. The organizational systems for performing work, improving processes, communication, involvement, leadership, management, ownership, accountability, reward, recognition, training, and education must keep pace with organizational and individual needs.

5. *Leaders develop others in the organization.* Other leaders are developed in the organization by current leaders who model, coach, and monitor the work force. The key long-term contribution of leaders to the organization comes in developing new leaders. In addition, people at all levels must be assisted in exercising their leadership and decision-making potential.

6. *Leaders give structure.* Leaders provide the organizational structure. They define the use of hierarchical, matrix, collegial, and team structures. They provide the support structure to ensure that TQM works.

Modern leader considerations

Leading by example (inspire)

Establishing a common focus

Acting to build and maintain teamwork

Driving productivity and quality improvements

Empowering others

Recognizing and rewarding appropriate performance (motivate)

Leaders make VICTORY happen

All the considerations of the modern leader are encompassed in the elements of VICTORY. Establishing a common focus is the vision. By acting to build and maintain teamwork the leader is involving people. Driving productivity and quality improvements equates to continuous improvement. Empowering others involves ownership. Recognizing and rewarding appropriate performance for motivation is a key element of VICTORY. Leading by example is how to display a yearning for success that inspires the organization to follow.

Vision

> *What you believe, you will receive.*

The vision provides a future state for the organization to strive to reach. The vision is one of the elements of a common focus.

Vision reflects the potential of the organization to establish long-term relationships with its customers and to participate and share in their continuous improvement. Vision is articulated first by top leadership, then worked through the organization. Vision is the outlook of the organization to be prosperous long into the future.

A vision must be developed by top management to indicate where the organization wants to go. The vision is what the organization sees as VICTORY. Since VICTORY varies by organization, each organization will have its own vision. Although top management creates the vision, it must be shared by everyone in the organization. The vision must also transcend the organization. Instilling the vision in others is accomplished by communication and example. The vision must be continually communicated through all means possible, and it should be repeatedly reinforced during all group gatherings. The status of the vision should be reported in information systems, newsletters, periodicals, and other organizational media. Further, and most importantly, the vision must be communicated by action. Management must set the example through leadership. The TQM philosophy and guiding principles can only be instituted in the organization by the personal actions of the leadership.

Leadership is essential to make the vision a reality. Leaders are not expected to have all the answers. Leaders simply guide the organization toward the vision. They must convey the importance of the individual in the organization and the role each contributor serves in making the vision a reality. This type of leadership is required throughout all levels of the organization. The leader may provide the guidance, means, and encouragement; but it is everyone in the organization who makes the vision happen.

Leaders establish a common focus

Chapter 5 details the action process with specific tools and techniques for establishing a common focus in an organization. The common focus consists of vision, mission, and values. Minimally, these common elements must be shared by everyone in the organization.

Involvement of everyone and everything

Unity provides strength.

The TQM framework requires the total involvement of everyone and everything in the organization. Everyone consists of the entire organization, including management/leadership, all the people in the organization, suppliers, customers, and teams dedicated to the ultimate goal of customer satisfaction. Everything comprises systems, equipment, and information. Figure 3.3 shows everyone and everything included in the TQM framework.

Management

Management through leadership ensures that the total organization is geared to total customer satisfaction. Managers give focus to the organization; design the processes that are used to perform the work; provide a framework where people can perform to the best of their capabilities; give everyone the means to do their specific process; and foster the development of a sense of pride and ownership of processes. They also empower the work force, invest in education and training, guide cooperation and teamwork, and motivate actions through rewards and recognition.

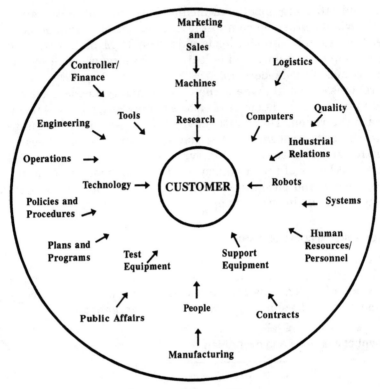

Figure 3.3   Involvement of everyone and everything.

Managers' considerations

Manages processes

Acts in a consistent manner

Nurtures workers' involvement

Allows participative problem solving and decision making

Gives power to workers to perform processes

Encourages pride of workmanship

Removes barriers to work performance

Supports efforts

Workers in the organization

All the workers in the organization must be empowered to perform their work with excellence. In a TQM framework, people are the most important resource. Therefore, they must be encouraged to be creative and innovative within all areas of their work. They must be allowed to make whatever changes are necessary within regulatory guidelines to perform the work and improve the system.

Workers' considerations

Work to perform and improve processes

Own their processes

Recommend improvements

Keep management informed

Encourage teamwork

Recognize individual and team contributions

Satisfy customers (internal and external)

Suppliers

Suppliers are important players in the TQM framework. They must be integrated into the TQM process, and they must understand the requirements of the organization. Further, the organization must develop a continuing relationship with its suppliers to ensure long-term customer satisfaction.

Supplier considerations

Survey your suppliers to see if they know your requirements.

Use suppliers that consistently meet your standards.

Partner with supplier for mutual advantage.

Provide incentives for suppliers.

Let suppliers participate in appropriate teams.

Integrate suppliers into internal processes.

Expect excellence from suppliers.

Reduce the number of suppliers.

Customers

Customers must also be integrated into the TQM process. The organization must weave the customers' needs and expectations into all its processes.

Customer considerations

Communicate with your customers.

Understand your customers' needs and expectations.

Survey your customers to determine if you are satisfying them.

Team with customers if possible.

Operate with a customer focus.

Manage your customer relationships.

Expect your customers to constantly increase demands.

Respect your customers.

Teams

The involvement of teams is critical to success in a Total Quality Management framework. Teams should be the primary organizational structure to accomplish critical organizational missions. Teams involve the internal organizational groups to include functional and especially multifunctional teams. In addition, suppliers and customers should also be participating in teams within the organization.

Team considerations

Together

Everyone

Achieves

More

Leaders act to build and maintain teamwork

Chapter 6 details the action process with specific tools and techniques for acting to build and maintain teamwork.

Include everything

The TQM management framework must include not only everyone but everything in the organization. Everything includes all the systems, processes, activities, jobs, tasks, capital, equipment, machines, vehicles, support equipment, facilities, tools, and computers. This management environment requires providing the proper means to perform the job. A proper balance of technology and people is essential.

The TQM management environment must include all systems, processes, activities, jobs, and tasks of the organization. A TQM system integrates all elements of the organizational environment and all functions of the organization. This organizationwide effort is absolutely essential. Many past improvement efforts focused on only one area of the organization such as manufacturing, design, or marketing. In addition, many policies and procedures did not allow the required improvements. Today, a total integrated improvement effort is the only way to success.

The organizational environment includes such items as communications, policies and procedures, training, rewards and recognition, benefits, accountability, evaluation, and marketing. The functions of the organization encompass engineering, marketing, human resources, manufacturing, finance, information systems, and logistics. In addition, all improvement efforts become elements of the TQM framework. These improvement efforts include the items shown under the TQM umbrella.

All the equipment must contribute to the attainment of the purpose of the organization. The equipment must assist people in performing processes while allowing them to add value through their ideas to the product service.

All information must be shared by everyone who has use for it. The information must show the current status of the organization as well as projections for the future. It must provide an accurate and comprehensive picture of all supplier requirements, internal process performance, and satisfaction of customer needs and expectations.

Information sharing is critical, as it shows management's commitment. Management must open up all information channels for a thorough information sharing to occur. It sometimes helps for management to translate traditional management information into a form that makes it easier for everyone to understand what they need to do.

Critical performance information must be predominately displayed to all people who need it. This type of information should be on charts that can be easily read and updated. When possible, the performance feedback should be constant and immediate. It must include all critical performance information in an organization. Performance feedback is essential to the TQM framework.

Continuous Improvement

> *There is a time for words and a time for action.*

Continuous improvement of product, processes, and people in an organization is essential, as shown in Fig. 3.4. Customer expectations drive product deliverables, which drives the people in the organization to improve their processes. Improved processes drive better product deliverables that exceed customer expectations. This process leads to the need for continuous improvement of product, people, and processes, which requires the establishment and maintenance of a disciplined continuous improvement system.

The continuous improvement system applies all the fundamental aspects of the TQM definition. First, people are not the problem. People are the solution. Almost all "root" causes of problems in an organization or variations in a process can be traced to the system or process itself. Therefore, the continuous improvement system uses

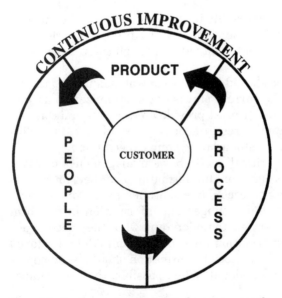

Figure 3.4 Continuous improvement of processes, people, and product.

people to focus on the system, process, issue, and problem; it does not look for fault in the people. Second, quantitative methods are the principle means to make decisions. Measurement is basic to all TQM activities in the entire organization. Third, the continuous improvement system with an appropriate improvement methodology is used to improve all material services supplied to an organization, all the processes within the organization, and the degree to which the needs of the customer are met, now and in the future.

Leaders drive continuous improvement

Leaders drive continuous improvement through the establishment and maintenance of a continuous improvement system. A disciplined continuous improvement methodology is required for victory, one that is used consistently throughout the organization. All the systems in an organization—the interdependent processes in the organization with a common purpose—must be involved.

The continuous improvement system focusing on process improvement must accomplish three objectives: first, it must bring processes under control; second, it must keep processes under control and make them capable; and lastly, it must continuously improve the processes aimed toward the best target value. This effort involves continuously eliminating waste, simplifying processes, and solving process problems and is a never-ending cycle.

Figure 3.5 shows a continuous improvement system. It is important to use the same continuous improvement system consistently throughout the organization. The system can be modified as needed for each specific application.

The continuous improvement system cycle involves five stages: (1) defining the focus, (2) determining improvement opportunities, (3) selecting improvement opportunity, (4) improving by using an improvement methodology, and (5) evaluating the results. A sixth stage can be added as a reminder—(6) do it again and again and again. This cycle is never-ending.

Stage 1: Define the focus. During this phase, the focus and priorities are determined. First, the overall focus must be established, understood, and supported. The focus must be determined by top leadership with input from all stakeholders. Once there is a focus, the specific mission is defined by the people responsible for the accomplishment.

Stage 2: Determine improvement opportunities. The next phase involves listing all improvement opportunities. It is important to obtain an understanding of the process at this stage. Customers, both internal and external, must be identified and their needs and expectations under-

Figure 3.5   Continuous improvement system.

stood. Suppliers must also be matched with requirements. Any potential problems should be identified at this time.

Stage 3: Select improvement opportunity.   Specific improvement opportunities are selected in this phase. Remember to focus on critical processes that have the greatest impact on customer satisfaction.

Stage 4: Improve using improvement methodology.   This phase uses a disciplined methodology to improve the process. This methodology is used to complete the mission, improve a process, and/or solve problems. There are many improvement methodologies.

Stage 5: Evaluate the results.   During this phase the improvements are evaluated against the impact on achieving the overall mission/vision of the organization.

Stage 6: Do it again and again and again.   This is a never-ending process. Everyone must continuously repeat the improvement cycle.

Training, Education, Coaching, Facilitating, Mentoring

The human resource is the main source of competitive advantage.

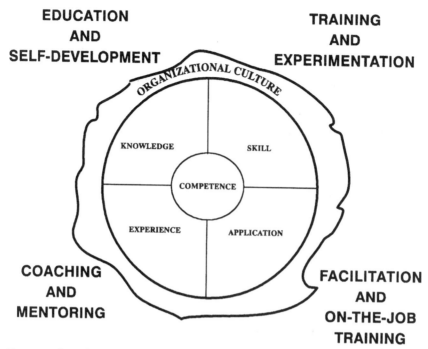

EDUCATION
AND
SELF-DEVELOPMENT

TRAINING
AND
EXPERIMENTATION

ORGANIZATIONAL CULTURE

KNOWLEDGE

SKILL

COMPETENCE

EXPERIENCE

APPLICATION

COACHING
AND
MENTORING

FACILITATION
AND
ON-THE-JOB
TRAINING

Figure 3.6 Learning process.

A people development system must be instituted. People development is a never-ending process for everyone in the organization. It is an investment that must be made, as it provides skills and knowledge—the ability to make it happen. Chapter 10 provides additional details for people development.

Figure 3.6 shows the learning process for competence in a specific organization. People must have knowledge, skills, application, and experience. These abilities are gained through a variety of methods such as education, training, coaching, facilitating, and so on. Each organization must create a learning organization by focusing on their specific organizational competence.

The education system must support the goals of the organization and the individual. Each organization must provide the opportunity for individual growth through education. In addition, each individual in the organization should be encouraged to pursue a life-long education process to foster future success for the organization and the individual.

In order to meet these needs, leaders must establish programs of education and training in the normal and routine structures of the workplace. In other words, education and training becomes one of the organization's core functions. The organization seeks training and education to improve the team's performance and understanding of customer issues.

Other key methods for people and organizational development, in addition to training and education, are coaching, facilitating, and mentoring. Coaching involves intensive application focused on specific results. Facilitating involves making the application easier by assisting with the conduct of a particular intervention geared to a specific process. Mentoring is the provision of an experienced person to offer guidance and support.

Specifically, the key skills that must be developed for a Total Quality Management framework include communication, especially listening; teamwork; conflict management; problem solving; consensus decision making; critical and systems thinking; understanding customer needs; and process improvement. In addition, leaders must cultivate skills of modern leadership as mentioned earlier to guide the organization to VICTORY.

Education

Education targets gaining knowledge. It provides an understanding of ideas, concepts, theories, history, background, situations, and so on. In TQM, education is important to develop the knowledge necessary for the specific application of TQM in a particular organization. It is also necessary to know the background, concepts, philosophies, principles, and teaching of the many masters of TQM. Education provides the knowledge to intelligently select from many alternatives for a specific situation.

Training

Training involves the application of knowledge, attitude, and skills to perform specific tasks. The training must be geared to the particular needs of the individual and the organization.

Training is geared toward developing and improving competence. The TQM framework requires everyone to gain additional capabilities to improve the process and perform the work. To do so, training is needed in TQM and job skills. Training in TQM philosophy, guiding principles, and tools and techniques is never-ending. Personal and team interaction skills must be continually refined. Specific job skills training must be provided and constantly updated to reflect the improved processes. All training must be geared to specific, clearly defined objectives. The training must be performed as closely as possible to the time when it is required. The abilities taught need to be immediately used by the trainee. Finally, all training requires reinforcement to ensure that VICTORY is achieved.

In TQM, training is essential not only for specific work operations but also for all the different skills of TQM. The organization needs to determine the specific competencies for success. This knowledge leads to specific attitude, knowledge, and skills for training. As a minimum, every organization pursuing TQM requires skills in understanding of customer needs, dealing with change, making process improvements, thinking critically, solving problems, communications, listening, and teamwork. The training needs to focus on development of the following:

- Modern leadership
- The TQM process
- Teams
- Team leaders
- Team members
- Facilitators
- Mentors
- Owners—managers, supervisors, associates, workers, etc.
- Other people with special TQM duties—e.g., quality department

Training considerations

Tell what is expected, the specific need for training, and what is in it for them.

Review frequently and differently to ensure retention.

Apply training to the real world.

Involve all participants and include life experiences.

Nurture individual differences in learning styles.

Coaching

Coaching involves intensive application focused on specific results. Coaching is geared to the particular needs and situation of an individual or team. The coach provides analysis, planning, instruction, application, and evaluation.

Coaching considerations

Coax answers.

Ownership remains with the individual or organization.

Assess the "real" situation.

Commend the positive, build on strengths, and eliminate weaknesses.

Help the individual or organization stretch expectations.

Facilitating

Facilitating involves assisting with the application of a specific process to make it easier to achieve desired outcomes. Facilitating helps learning by doing. Facilitators are important people in a TQM organization support structure. Although they are not team members, they help the team reach its mission.

Facilitating considerations

Focus on shared purpose.

Assist with the application of principles, methodology, tools, and techniques.

Coach as appropriate before and after meetings.

Interact during sessions when appropriate and necessary.

Let the people add the content.

Intervene to help manage conflict.

Train specific skills, tools, and techniques.

Assess performance.

Take the initiative to make things right without imposing your own will.

Encourage open communication and active listening.

Mentoring

Mentoring is the provision of an experienced person to offer guidance and support. In TQM, mentors are particularly useful in the role of supporters of team leaders and teams who interact with management. As such, they need to have credibility and influence in the organization. They support the team leader by offering a listening ear.

They also provide welcome advice on handling issues that involve internal politics. Frequently, a mentor can assist the team in introducing and implementing a critical TQM initiative that is counter to the present organization culture.

Mentoring considerations

Master who is a role model for others

Enlightens on organizational culture based on experience

Nurtures the progress of the team leader and team

Teaches life or organizational lessons

Orients toward what is best for the organization

Referees organizational issues between team leader or team and organization

Ownership

> *People who own something will take care of it.*

Although all the elements of VICTORY are essential to success, ownership is one element that is often overlooked by many organizations. People in the organization must feel that they own TQM, systems, processes, and actual work.

The ownership by the organization of its own TQM process is key to TQM success. One of the causes of failure of many TQM efforts is the feeling of "not invented here" that comes from the organization simply adopting someone else's program. Ownership provides the support and commitment necessary to maintain TQM over the long term. If you own the process, you take responsibility to ensure that it is a success. Although every organization needs a process to follow, most successful TQM processes are the result of ownership by the specific organization. Many organizations even have their own name for their particular version of TQM, essentially saying to the world "this is mine."

Ownership of anything, including the TQM process, is important to get results in the workplace. It implies the ability to perform and improve systems, processes, and the work itself and it involves encouraging and empowering people to create ideas and make decisions. Ownership is important to ensure pride of workmanship. It implies responsibility, authority, and resources, which encompass the boundaries of empowerment. Everyone in the organization, including top leadership and all workers, must be given ownership.

Besides individual ownership, team ownership is equally important in a TQM framework. If everyone owns their work, the entire organization can work with pride and commitment toward satisfying the customer.

Leaders empower others

Ownership starts with empowerment. What you have the power to do, you own. The leader accomplishes the common purpose of the organization by empowering others. Empowerment means that all individuals and teams in the organization have the authority to do what is necessary to perform and improve their work. Empowerment involves having the responsibility, authority, and resources to do whatever is required to satisfy the customer and achieve the mission within defined boundaries. The empowerment process involves the gradual shifting of responsibility, authority, and resources to people in the organization who are performing and improving the work. As the leader empowers others, the whole organization improves.

Empowerment considerations

Encourages ownership of work

Makes everyone an intrapreneur/entrepreneur

Promotes relationships with owners, suppliers, customers

Opens opportunities

Works toward a common purpose

Encourages win-win outcomes

Recognizes and expands boundaries

Empowerment action process

1. Define the traditional boundaries of authority.
   - Mission
   - Vision
   - Process description
   - The process beginning and end
   - Expected results with performance measures (e.g., goals, metrics)
   - Customer(s)

- Resources (people, time, equipment, capital, and money)
- Key responsibilities

2. Identify strategies to empower others in the organization and/or team.

3. Develop a plan to empower others.

4. Support the empowerment.

- Recognize and reward achievements.
- Provide mentoring, coaching, facilitating, training, and education.
- Build on lessons learned.

Recognition and Rewards

If there is a reward, people will seek it.

Recognition and rewards must be instituted to support the TQM framework. Although recognition and rewards are elements of any organization, the TQM framework requires changes to the usual recognition and rewards systems. The recognition and rewards system must foster appropriate behaviors and extraordinary actions to improve the organization.

Recognition and rewards are shown in Fig. 3.7. A reward is given for performance of some specified action. Rewards can be extrinsic, like compensation, promotion, and benefits, or they can be intrinsic, like feelings of accomplishment, improved self-esteem, personal growth, or sense of belonging.

Recognition is given for special or additional efforts. It takes the form of praise or a celebration. Praise should be the normal method to reinforce the right behavior. Celebrations can be individual or group. The recognition and rewards systems of the organization must foster the TQM philosophy and guiding principles. They must constantly and immediately reinforce leadership, teamwork, individual contributions, continuous improvement, and customer satisfaction behavior.

The TQM framework requires people to add new responsibilities. The reward systems must recognize this achievement with new rewards. Any new reward system must be equitable and just. Further, it should include an appropriate combination of extrinsic and intrinsic rewards.

Leaders motivate others by recognition and rewards

The leader accomplishes the common purpose through the performance of others. The behavior of an individual or teams must be mo-

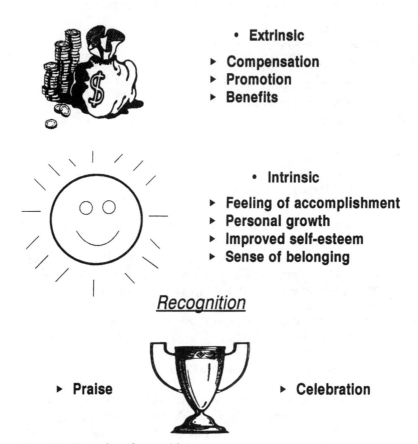

Reward

- **Extrinsic**
  - ► **Compensation**
  - ► **Promotion**
  - ► **Benefits**

- **Intrinsic**
  - ► **Feeling of accomplishment**
  - ► **Personal growth**
  - ► **Improved self-esteem**
  - ► **Sense of belonging**

Recognition

► **Praise**   ► **Celebration**

Figure 3.7 Rewards and recognitions.

tivated to selectively direct energy toward the common purpose. Motivation is an essential component of getting others to perform. Motivation plus ability are the two key elements of performance. A person is motivated to do what he or she is rewarded to do. The leader needs to motivate each individual in the organization to pursue and accomplish the common purpose of the organization.

The challenge for today's leader is to motivate by creating opportunities for each individual in the organization. In addition, the leader needs to provide both traditional and nontraditional rewards.

Motivation considerations

Make sure the common purpose is shared.

Orient and integrate new members.

Think and speak "we."

Institute appropriate rewards and recognition.

Value individual contributions.

Avoid frequent changes of members.

Take time to exchange greetings.

Encourage a sense of belonging.

Motivation action process

1. Select a champion for developing a recognition and reward system.
2. Start with a simple system that meets the following guidelines:

Recognize both individuals and groups.

Have rewards that each individual can choose for her- or himself.

Add special recognition and rewards as appropriate.

Recognize and reward appropriate behaviors and actions.

Motivate everyone toward performance. This is the system-driven purpose.

Have rewards separate from performance appraisals, merit, or annual review. These systems are barriers to the TQM process.

Provide a formal system, but allow flexibility within the system.

Allow the system room to change, but keep some highly sought after rewards constant.

Provide a variety of options for recognition and rewards to allow for the ever-changing needs of individuals and the organization.

Permit recognition to be given by anyone inside or outside of the organization.

Give recognition and rewards regularly, not only once a year.

Have management give rewards.

Initially, keep rewards separate from the compensation system.

Keep it fair, simple, and fun.

Don't consider cost an issue—many things can be done at little or no cost. Rule of thumb: an organization should invest as much in its own people's motivation as it does on motivating customers.

3. Initiate a recognition and rewards system in the organization.

4. Evaluate the system every 30 days to update, reinforce, and revitalize.

5. Eventually, the rewards system should be integrated with the compensation system with some type of pay for performance and/or value or a profit-sharing scheme, as appropriate, to achieve the organization's objectives.

Yearning for success

> You must want it to get it.

Leaders must have the intense desire to win. They must commit to long-term support and must be willing to make an investment of their personal time and the organization's resources. They must understand that although some results will be quickly realized, permanent changes will take many years. Commitment to this long-term objective involves a number of different actions. First, leaders must set the example by displaying the expected behaviors day after day. Second, leaders must provide the resources. Third, a support system must provide the direction, guidance, and support to the overall TQM effort. Leadership must be an active, highly visible participant in all aspects of the TQM process.

Leaders must have the discipline to make this long-term commitment for the future of the organization. They must thoroughly understand the TQM process. In addition, the TQM philosophy and guiding principles must be constantly and consistently applied throughout the organization. This effort requires another commitment by leadership to devote personal attention to its implementation.

Chapter 11 provides additional guidance for yearning for success.

Leaders achieve success by setting the example

> People do what they see the leaders doing.

As a leader, one of your most important roles is to set the example for others. By setting an example for others to follow, the leader creates and maintains an organizational environment conducive to success.

The example the leader sets in an organization has a great influence on the organizational environment. The leader's characteristics, and especially the leader's behaviors, have a great impact on the success of the organization.

The modern leader must continuously demonstrate behaviors she or he wants from others in the organization. Leaders must show commitment and support to guide the organization. They cannot just pay lip service or practice deception. Leaders must display commitment and support through appropriate words and actions. They must "talk the talk," "walk the walk," and "walk the talk."

The leader's example forms the foundation for all other actions in the organization. If the leader has trust, others will trust. If he displays ethics and integrity, others will act ethically and with integrity. If she has constructive relationships, others will have constructive relationships. If he or she welcomes change and risk, others will champion change and take risks. In sum, the leader can expect to get back what he or she gives out.

Leading by example considerations

Encourage trust through open and honest communication.

X-ray your power and influence on the organization.

Act to make things good, better, best, excellent.

Maintain ethics and integrity.

Pursue constructive relationships.

Listen to others' points of view.

Encourage change, innovation, and risk taking.

4

Installing, Reinstalling, or Revitalizing Total Quality Management

TQM requires a disciplined ongoing action process of analysis, design, development, implementation and evaluation.

> *There is more than one way to get to the same destination, but it always helps to use a map.*

Many organizations are either looking to install TQM for the first time, reinstall a failed TQM process, or revitalize a stalled or mature TQM effort. TQM is a never-ending process. It starts with a foundation of basic business ethics that must underline all TQM efforts. Integrity, ethics, and trust form a solid foundation. Open and honest communication channels further the foundation for a TQM environment. Adding to this foundation is the TQM framework. The elements of VICTORY-C guided by leadership add in the creation and maintenance of the TQM environment. The TQM environment allows the TQM process to be built. The TQM process requires a systematic, integrated, consistent, organizationwide approach. It involves a lot of hard work, starting at the top of and extending throughout the entire organization. The organization must be turned inside out, examined, and changed as appropriate for VICTORY to be achieved in that specific environment. This goal cannot be accomplished overnight. It must be created through many small, continuous successes over time.

Each organization must determine what specific changes are necessary to create and maintain the TQM process to achieve victory. Within each organization, a distinct operating environment exists which has a great impact on the performance of the organization. The necessary changes may require a new or redesigned reward system; additional members; a different organizational culture; a change of management or leadership; involvement of more employees in decision making; more ethical behavior; development of supplier partnerships; establishment of customer relationships; increased integration of functions; a modification of the organizational structure; development of trust among employees, suppliers, and customers; establishment of a sense of belonging; development of (self) discipline; installation of measurements; creation of a new organizational way of life; and so on. The TQM process is created and maintained through a never-ending, continuous-loop action process.

There are many road maps to VICTORY. All the roads must be examined to find the best path for each specific organization. There can be no quick fix, no shortcuts, and no magical formula. TQM requires many years of dedicated, hard work to reap the many benefits of this never-ending organizational process.

Although there are many roads to VICTORY, most organizations develop their own path based on the many proven paths. For instance, the VICTORY-C model is one path. Other paths, including those given by specific masters of quality or that use certain criteria like the Malcolm Baldrige National Quality Award, provide a variety of alternatives for an organization to follow. Most paths have a proven success in some organization. In addition, any proven path tailored to a specific situation should lead to success. Therefore, it is important to pick or design a path that would be most appropriate for your organization to get it to the desired destination. It is critical to own and stick to the chosen path while looking for continuous improvement as situations demand.

This book provides a practical approach that can be adapted to any organization for VICTORY. This process can be used for any organization looking to install TQM for the first time, to reenergize a low-performing or failed TQM effort, or even to boost a successful TQM process. In fact, this process should be repeated in a never-ending cycle to adapt TQM to ever-changing conditions.

Before beginning the actual process, the organization must prepare by performing a needs and feasibility assessment to decide on using Total Quality Management. Once the decision to use TQM is made, the organization can begin the journey through the TQM action process.

TQM Action Process

The TQM installation action process consists of the following:

Get ready = ASSESS

Get set = DESIGN

Aim = DEVELOP

Fire = IMPLEMENT

Reaim = EVALUATE

The above list gives two names for the same processes. For instance, get set and design are all listed in the second line. The first is a common descriptive name. The second is the action process name. Each of the names can be used interchangeably. The first name is to remind practitioners of the sequence of the process. It is critical to use discipline to follow this process exactly. Many organizations have failed because they fire before they aim or do not get ready or set first. Many times practitioners try to fire before getting ready, getting set, or aiming. If you don't do these processes before firing, you never know what you will hit. The second name is the more formal name for the processes. These are the action processes each organization must continuously apply to make the installation of TQM successful. The focus is on "action." Figure 4.1 graphically shows the common descriptive names and the TQM installation action processes.

The processes in the TQM installation process flow over a period of time. They do not need to be accomplished one after another. To expedite the overall process, each step overlaps the previous one. The amount of time for each individual process and the overall process varies by the organization. In addition, the action processes must be continuously repeated. The organization uses it once for installation, repeats it for the operation phase, and repeats it again for institution, and repeats it again during the next phase in a never-ending cycle.

Figure 4.1 also shows the output of each of the processes. The preparation process provides the need to pursue TQM. From here, the assessment process gives the organization information for the other processes. These can take any format as deemed appropriate for the organization. The organization assessment report card is a generic name for the output of this process. This is the input to the design process where the output is TQM plans. The next process takes the development actions from the TQM plans to create the TQM system. This TQM system as implemented becomes the TQM process. This TQM process is evaluated for continuous improvement.

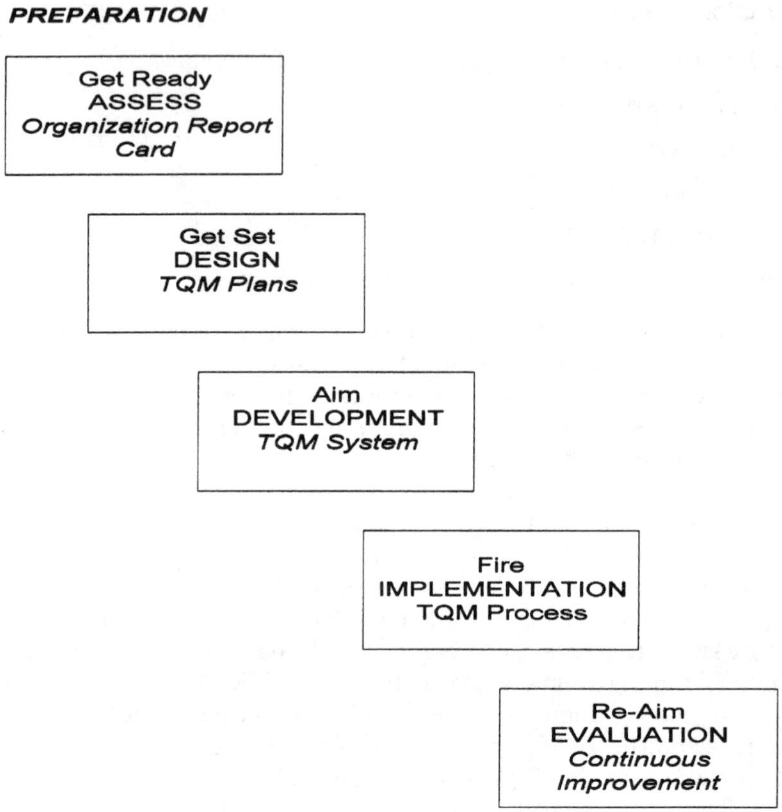

PREPARATION

```
┌──────────────────────┐
│     Get Ready        │
│     ASSESS           │
│ Organization Report  │
│      Card            │
└──────────────────────┘
```

```
┌──────────────────────┐
│     Get Set          │
│     DESIGN           │
│    TQM Plans         │
│                      │
└──────────────────────┘
```

```
┌──────────────────────┐
│      Aim             │
│   DEVELOPMENT        │
│    TQM System        │
│                      │
└──────────────────────┘
```

```
┌──────────────────────┐
│      Fire            │
│  IMPLEMENTATION      │
│   TQM Process        │
│                      │
└──────────────────────┘
```

```
┌──────────────────────┐
│     Re-Aim           │
│   EVALUATION         │
│   Continuous         │
│   Improvement        │
└──────────────────────┘
```

Figure 4.1   TQM Action Process.

Preparation

> *The better the preparation, the greater the results.*

The input to the preparation process is the perceived need for TQM. The preparation process assesses this need and feasibility for the specific organization. The output of the preparation process is a decision to pursue TQM. Preparation starts the TQM project. The initial installation of TQM in an organization should be viewed as a project. As such, project management techniques should be used to ensure a successful project. In this case, the TQM project is a success when TQM is instituted into the organization. In other words, TQM becomes the way of life for the organization. Therefore, it is necessary to constantly balance the requirement to synthesize TQM into the everyday organization with the need to take time separate from day-to-day activities to create the TQM process.

Every project should start with an agreement on the following:

1. Focus (what)

2. Organization (who)

3. Process (how)

Focus

> *What do you want to do?*

The initial focus for installing TQM starts with the action process (see Chap. 2). Someone in the organization, preferably a top-management representative, must have the answer to the following questions:

- Does the organization have a foundation for TQM?
- Can the organization deal with the change necessary for TQM?
- Is TQM the right approach for the organization?
- Is the organization ready for TQM?

Responses to these questions should be clarified in a focus statement. If the organization cannot define a business reason for TQM, then they do not need to go further. However, most organizations find TQM an imperative for survival today and success in the future. This fact is evident in the number of companies, organizations, governments, and even countries that are pursuing the TQM path. It is clear that TQM is the right path. The focus statement should be a simple, concise sentence of no more than 25 words. The focus statement must convey a sense of urgency or purpose for TQM in the specific organization.

Organization

> *Who is going to do it?*

The TQM organization starts with formation of a team. The composition of the team depends on top-management support and the most likely approach to implementation of TQM. Ideally, this team should be the foundation for the executive steering committee/group/team, quality council, focus team, or core team. These are the names of top-management teams formed by virtually every organization that has successfully implemented TQM. This team is the project management team. It targets TQM to focus on the organizational strategy. This team provides identity, structure, and legitimacy to the TQM effort. It must be eventually formed for implementation of TQM. As such, it would include the top manager and key players in the organization needed for critical mass. Ideally, the leader of this team should be a top leader in the organization with direct operational responsibilities. If the top leader cannot personally lead the team, and in many organizations this arrangement is not possible, especially in large organizations, he or she needs at least to be the mentor to this team. It is critical to future success to have the top leader actively involved in certain critical activities with the team, especially in the beginning. In all cases, this task should never be delegated to the quality assurance manager. The quality manager is an essential member of the initial team, but TQM requires a bigger picture than just product or service quality. In addition, this type of delegation leads to TQM becoming a "quality" thing instead of an "organizational" strategy for success.

This team should be supported by a qualified TQM coach and facilitator. The role of the coach and facilitator is to assist the top leader and team in getting off to the right start. This support resource provides expertise not available in the organization to ease the transition to TQM. In addition, this resource provides a different view of the organization and he or she must be able to tell it like it is without regard to internal politics or personal job security.

This team forms the initial TQM organizational structure that initially runs in parallel and later becomes integrated with the normal organizational structure. It is recommended that this initial team resemble the normal organizational structure as much as possible with the addition of a few additional stakeholders, for example the union and workers. Eventually, the TQM structure must be synthesized into a new way of organizational life.

If integration of TQM within the organization is not possible during preparation, a TQM study team can be established with the support, or at least approval, of top management and/or steering group with a mission to provide a specific recommendation for TQM. This team could comprise key players or champions throughout the organization who are considering implementation. It is also recommended that representatives of suppliers, customers, and union, if appropriate, be invited to participate.

Again, if there is not enough support to form a team to look into the feasibility of TQM, champions in the organization must get the required support before pursuing any TQM activity on an organization-wide basis. Champions of TQM can initiate in their own areas some of the TQM processes, methods, tools, and techniques as "best" management practices. Once they have achieved success, others in the organization will want to follow the champion's success.

In order for TQM to succeed organizationwide, top-leadership support and commitment is necessary. Although TQM can have an impact in an organization even without top-leadership support, it is better to have top-leadership support as early as possible. After all, the main reason for using TQM is to have organizational success. All top leaders have this desire, and everyone wants to be part of a winner.

Top-leadership commitment

Top-leadership commitment is absolutely necessary for the long-term success of any organization. Top leadership determines all the elements of VICTORY, as it has a major influence on the methods used to conduct the organization's business, organizational culture, and individual and organizational performance. Therefore, top leadership must believe in the use of the TQM process as the way to conduct business for overall organizational success—and be involved with and support it. Sometimes, top leadership by word or action can change an organization overnight where an organizational development intervention could take many years to achieve the same change. This fact does not mean that top leadership can simply announce "TQM is the way of life" and it will happen. It does mean that top leadership support is necessary for success.

Which top-management support is required?

Top management often is defined differently by different organizations. What is considered "top management" for the purpose of implementing TQM depends on the organization. Ideally, the chief executive officer, president, general manager, head of an agency, or highest officer in an organization would be considered top management for an organization. Obviously, active involvement by top management would lead to an improved probability of success for the entire company, division, agency, or organization.

However, the total-quality process can be applied anywhere in an organization, from the shop floor, to functional organizations, to staff areas. Wherever it is applied, the TQM process can be implemented within that particular area of ownership. Each group or division needs to use the process of getting started to carefully define the boundaries of its control while seeking to continuously gain the support of others in the organization. At a minimum, that area needs the support and commitment of the next highest level of management in the organization.

Remember, the full benefits of the TQM process are extremely difficult to achieve without the support and commitment of the top leaders in an organization. However, the TQM process can be used for success in any organization.

Getting top-management support

Since top-management support is essential to success, sometimes TQM champions must actively work at getting it. The TQM process must focus on the long-term implementation within the entire organization. Champions must address the specific strategy for getting top-management support. The strategy chosen will depend on the organization culture and the individual top manager. The actions used to gain the support should be geared to the specific desired outcomes. Either individual or combined actions can be used over a period of time. Some specific actions that have been successful in other organizations include

- Talk to other top managers that experience success through TQM.
- Attend education and training session.
- Use selected films, books, or articles.
- Take on-site visits to other organizations.
- Present examples of successful TQM implementation at similar organizations.
- Review case studies.
- Participate in conferences and seminars.
- Perform a successful pilot project in the organization.
- Solve a difficult problem using the TQM process.
- Involve "champions" to share success.
- Document and measure improvement results.
- Relate TQM to goals of top management.
- Show how use of TQM already helps the organization.
- Win the support of "key" players in the organization first.
- Determine outcome of strategic planning for the organization so TQM process can proceed.
- Provide an environment for the top manager to discover the need for TQM.
- Make TQM the top manager's own idea.
- Hire outside assistance.
- Customer acquires TQM from suppliers.

Process

> *How are you going to do it?*

This book represents one process for pursuing TQM. It provides a process for tailoring the content to any organization. There are many other processes for pursuing TQM, including the Deming Method, the Juran Trilogy, and the GOAL/QPC model. Each organization must do research to determine their specific road map for victory. When one embarks on the TQM journey, it is important to use a disciplined, systematic, integrated, consistent, organizationwide approach to determine the specific road map for VICTORY. Most improvement efforts are derived from a combination of many sources. Therefore, understand as much as possible about the various options. The approach can be adopted as is, modified, or created. Whichever approach is used, it must be embraced or owned by the entire organization. If the approach is understood, advocated, supported, and owned by the organization, the chances of VICTORY will be increased.

As the practitioner goes through this book, he or she must continuously decide, based upon his or her own situation, what specific content to include or actions to take. You should use the process exactly as prescribed. If you decide to use the process outlined in this book, you need to determine within each process the specific content as follows:

1. Apply as is
2. Modify
3. Does not apply

Do you need help?

> *Beware of the wolf in sheep's clothing.*

As you prepare for and install the TQM process, you will need assistance. As a minimum, the organization will need to learn as much as possible about TQM. There are many opportunities for self-development, including books, reports, research papers, and even on-line services, In addition, many organizations that have experience with TQM are willing to share information with others. These avenues should be pursued by every organization in preparation for TQM.

TQM, although simplified in this book, is a difficult and complex activity in many organizations. For a small organization, use of this book may be sufficient to get significant results from TQM. For large and medium organizations, some outside assistance is almost always required for success.

Many organizations have internal TQM or quality expertise. The organization should take advantage of its knowledge and skills to build critical mass within the organization. However, these internal people will have to deal with the politics of the organization. They may not always be able to tell it like it is to the top management in the organization.

There are many experts to help guide an organization to its destination. Most of these people focus on certain aspects of TQM. For instance, the quality consultants provide various approaches to achieve product or service quality or sometimes only to comply with ISO-9000 Standards. Other experts give advice on changing the organizational environment. Still others give guidance on teamwork, leadership, improvement processes, manufacturing resource planning, concurrent engineering, statistical process control, and so on. Further, some organizations will assist only in implementation. These consultants or organizations can provide assistance on their specific expertise. But beware: some consultants proclaim that the solution to every problem happens to be their specific solution. What solution is needed depends on the organization and situation, and it can vary over time. The means must always focus on the specific ends of the organization.

In most cases, an organization should seek assistance from a TQM process expert who either can do it all or has access to the required expertise, as required. A word of caution: Some organizations only seek a TQM expert in their specific field or industry. In most cases, when an organization pursues TQM, a TQM process expert is all that is necessary. The organization should be the expert in the field or industry. It can then keep the all-important ownership of the results for its TQM function. This does not mean that an organization should not use a TQM process expert who knows the field or industry. It also does not mean that an organization should only use a person who is just a TQM process expert. It means that the organization needs to weigh the advantages and disadvantages of both types of experts for their specific situation before making any selection. For the organization's TQM success, the TQM process expert is the most needed assistance.

The following are some of the elements to look for when seeking TQM assistance:

- Does the consultant or organization offer more than one solution to your problem, or is its "product" always the solution to every problem?

- Are there "different strokes for different folks," or does one size fit all?

- Does the consultant or organization use a systems approach, or are they selling one specific approach?

- Can the consultant or organization perform many interventions such as coaching, facilitating, and training, or do they just give recommendations and disappear?

- Can the consultant or organization provide enough attention to you, or will you get either multiple consultants or limited time?

- Does the consultant have hands-on experience, or has he or she just gone to glamour school?

- Do you have the rapport with the consultant or organization to build a relationship, or is the situation strictly a business deal?

- Does the consultant or organization practice what they preach?

A word to the wise

> *If it sounds too good to be true, it probably is.*

The TQM process requires a systematic, integrated, consistent, organizationwide approach. It requires lots of hard work, starting at the top of the organization and extending through the entire organization. There are no quick fixes, magic wands, or shortcuts. The organization must be turned inside out, examined, and changed as appropriate for VICTORY in its specific economic environment. This goal cannot be accomplished overnight. It must be created through many small, continuous successes over time.

Many people think that all that is required to perform TQM is to do some reading, talk to another organization, attend a college course, or participate in a seminar or workshop. This conception, frankly, is not true. The TQM development effort for a successful organization and individual requires a long time. TQM does not just happen. The president cannot simply announce it and it will happen. People cannot take a class and become instant experts. A successful TQM effort can take from one to seven years to institutionalize. Therefore, investing in long-term outside support, as soon as possible, with the ultimate goal of developing internal capabilities is critical. Even when you develop internal capabilities, TQM is a never-ending, ever-changing approach that may require "maintenance" support from time to time. For organizations of 25 or more people, an outside TQM coach is recommended for at least one year, with larger organizations needing more time with available support for as long as may be needed for success. For organizations of less than 25 people, a TQM coach may only be required for six to nine months with long-term support.

Preparing for TQM Action Process

1. Assess current TQM foundation, including communication.
2. Evaluate the organization's capacity to deal with change.
3. Determine if TQM is the right approach for the organization.
4. Decide if the organization is ready for TQM.
5. Decide on a focus statement for TQM.
6. Form executive/steering/core team.
7. Agree on specific process for TQM.
8. Learn as much as possible about TQM.
9. Get outside assistance.
10. Make the decision to pursue TQM.

Assess

> *The organization that takes time for intelligence gathering beats the competition.*

Focus: Determine the present state of the organization.

Organization: A top-management team—executive team, steering team, core team, or strategic TQM/quality council—consisting of major stakeholders in the organization. This team will get assistance from internal or external resources, as necessary. Resources are required for coaching top executive(s) and educating, training, and facilitating the team.

Input: Decision to pursue TQM

Process: Assess organization focusing on achieving total customer satisfaction

Output: Organization assessment report card

Assessment action process steps

1. Establish a focus statement for the team.
2. Form the team and get commitment.

3. Understand and define specific assessment action process steps.

4. Perform assessments of

Customer needs and expectations

Organizational culture

Internal processes

Competition

5. Compile an organizational assessment report card.

Assess the organization

The assessment of the organization starts with the focus on total customer satisfaction, which requires the organization to know its customers. By knowing its customers, the organization focuses the development of relationships to keep and gain customers both internal and external. In addition, the organization must know itself, its products and/or services, and competition. By knowing itself, the organization understands what it can do to satisfy its customers. By knowing its products and/or services, the organization can position the deliverable to maximize total customer satisfaction. By knowing the competition, the organization can establish targets for gaining advantages in the marketplace. Figure 4.2 shows the elements that must be assessed in order to achieve total customer satisfaction.

Know your customers

Chapter 3 describes the process for focusing on customer satisfaction.

Know yourself

The organization must know itself to achieve total customer satisfaction. In the process of knowing yourself, the organization looks inward to its organizational culture and internal processes. In this process, the organization is examining its own way of doing business.

It must examine its culture to discover the true nature of the organization. The organizational culture includes language, behavioral norms, ceremonies, informal and formal social and work process, organizational management style, reputation, image, philosophies, values, attitudes, beliefs, assumptions, and traditions. The organizational culture has a major impact on successful performance of TQM.

Figure 4.2 Elements for assessment.

Culture considerations

Organizational culture is important because it affects the following:

Communications, internal and external

Unity for cooperation and teamwork

Leadership and followship styles

Time and urgency factors

Uniqueness in the marketplace

Relationships

Effectiveness, efficiency, productivity, and quality

In addition to its specific culture, the organization must know its internal processes. Internal processes focus on getting the work done with desired internal quality while targeting total customer satisfaction. Each customer in the organization must be satisfied. However, unlike the external customer, the internal customer may or may not be a person. The customer could be a person receiving the output of the job or the next process, the next task, the next activity, the next job, the next machine, or the next piece of equipment. To satisfy internal customers, all the processes in the organization must be understood, measured, and analyzed to determine existing performance.

The success of TQM depends in large part on an organization's ability and willingness to perform TQM. The organizational culture and internal processes both play a critical role in the willingness and ability of the organization.

Know your product and/or service (deliverables)

To achieve total customer satisfaction, the organization must know all there is to know about its product, including all its aspects. The product is an output of a process that is provided to a customer (internal/external) and includes goods, services, information, and so on. The product is all aspects contributing to customer satisfaction, including product quality, reliability, maintainability, availability, customer service, support services, supply support, support equipment, training, delivery, billing, and marketing. Again, every one of these elements of the product and/or service must focus on customer satisfaction. The goods or services may be the best in the marketplace, but it is the entire product that contributes to total customer satisfaction. If the product does not provide total customer satisfaction, the customer will not be satisfied and most likely will go elsewhere to find a product that gives total customer satisfaction.

Total Product Concept

The Total Product Concept from *The Marketing Imagination* by Theodore Levitt of Harvard Business School provides an insight into the range of possibilities for a product to satisfy needs and expectations. Looking at Levitt's circle in Fig. 4.3, the generic product is the basic item. The expected product is the customer's minimal expectations which includes the generic product. The augmented product offers more than what the customer has become accustomed to expect. A potential product is anything that can be used to attract and hold customers beyond the augmented product. The Total Product Concept is important when targeting the product and/or service required to achieve total customer satisfaction.

When striving for total customer satisfaction, the product and/or service, at a minimum, must be comparable to the product offered by the competition. Obviously, a competitive advantage is gained by improving the product. The product or service may be differentiated by raising the level of customer satisfaction. A thorough analysis should always accompany any targeting of a product for customer satisfaction.

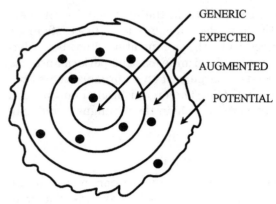

GENERIC

EXPECTED

AUGMENTED

POTENTIAL

Figure 4.3 Levitt's circle.

The goal is always to optimize customer satisfaction with resources. Raising the level too far beyond the current range of customer satisfaction is risky because of two factors. First, the cost factor could impact perceived value. Second, the customer may not be ready for the enhanced product or service. In both cases, the product or service may not be sold because it did not satisfy the customer.

Know the Competition

The organization must know the competition to establish targets for its products and/or services and internal improvement efforts. The organization must establish targets for products and/or services in relation to its competition. Typically organizations compete in the major areas of technology, cost, product quality, service quality, and time. Frequently, product competition progresses from one area to the next. It starts with technology, then cost, then product quality, then quality service, and then time or some other area.

For instance, when the television first was marketed, technology was the foremost satisfier with the customer. Once the technology was readily available and television become more desirous to customers, price became the most important element. As the television became commonplace, customer satisfaction was judged through product quality; and quality service was the differentiator. Today, television competes on all levels at the same time.

The organization must know in which major area of competition its product and services are competing. This determination will show the organization where to target its product and services. In addition, the organization should always attempt to achieve the competitive advantage. Therefore, it is important to look for ways to differentiate within the major areas of competition.

The organization must know where it is in relation to the competition to establish targets for internal improvement efforts. To determine its position in the competitive market, the organization should benchmark itself against its top competitor and the best in the field. Once benchmarks are determined, the organization can establish internal targets for improvement efforts.

Assessment Tools and Techniques

The scope and depth of assessment depends on the specific organization, and the list of tools for assessment is bountiful. There is no shortage of assessment assistance. Again, the organization should only focus on selecting the tools and techniques that would provide the organization the greatest benefit with consideration of cost and time.

Tools and techniques include some of the traditional tools such as surveys, interviews, management style profiles, and the Myers-Briggs Type Indicator. In addition, various award criteria or standards could be used for assessment, such as

- Malcolm Baldrige National Quality Award
- President's Award for Quality and Productivity Improvement
- Deming Prize
- Quality Improvement Prototype Award
- The Quality and Productivity Self-Assessment Guide for Defense Organizations
- ISO-9000 or ANSI/ASQC Q-9000 series of standards

Some of the tools and techniques useful for assessment in this book are:

- VICTORY-C assessment
- Teamwork assessment
- Malcolm Baldrige National Quality Award checklist
- Quality function deployment
- Benchmarking
- Metrics

Compile an organizational assessment report card

The organizational assessment report card or "state of the organization" to achieve total customer satisfaction provides the baseline information for the remaining processes. In other TQM action processes, this information is used, updated, clarified, and modified as appropriate. As a minimum, the organizational assessment report card should include:

- Organizational assessment against overall TQM desired outcomes, criteria, or standard—e.g., VICTORY-C, Malcolm Baldrige National Quality Award, and ISO-9000
- Critical processes performance indicators
- Deliverables place in the Total Product Concept
- Customer needs and expectations evaluation

Design

> *Plan the TQM system to meet the needs of the organization.*

Focus: Create TQM plans.

Organization: A top management team—executive team, steering team, core team, or strategic TQM/quality council—consisting of major stakeholders in the organization. This team will get assistance from internal or external resources, as necessary. Resources are required for coaching top executive(s) and educating, training, and facilitating the team. In addition, the organization many require support assistance in preparing, conducting, and evaluating various assessments.

Input: Organization assessment report

Process: Prepare TQM planning documents

Output: TQM plans

Design action process steps

During the design process, the organization plans its own TQM system appropriate for organizational success. The TQM plans are the result of the design process. Each organization decides on the scope and breadth of its own TQM plans. During design, the strategic quality plan gives the focus, policy, and objectives for the organization. The quality system plan provides the requirements for the quality system for improving customer satisfaction while managing costs of internal processes. Process plan provides an understanding of processes.

Typical design action process steps are

1. Create a common focus

   ▪ Vision
   ▪ Mission
   ▪ Values

2. Establish quality policy
3. Prepare the strategic quality plan
4. Prepare TQM system plans
5. Perform process planning
6. Document TQM plans

Chapter 5 describes each of these steps in detail.

TQM planning is action planning

A TQM system requires action planning. Action planning differs from traditional planning in that action planning is a living plan for focusing the organization's actions. Before we describe action planning, the planning process must be understood. The top portion of Fig. 4.4 provides an overview of the typical planning flow. Planning is both top-down and bottom-up. The plan involves only certain people at each level. The planning process starts with top management defining the objectives. Next, top management and middle management determine the specific goals to support the objectives. Then, middle management develops a strategy to meet the objectives and goals. Finally, middle management and all other employees determine and perform the tactics and operations necessary to make it all work.

Action planning is shown in the bottom portion of Fig. 4.4. Everyone is part of the action planning process. Action planning is dynamic. With action planning, top leadership still determines the vision and objectives but the entire organization is involved. The objectives, goals, tactics, and operations are integrated into the organization's way of life. Each phase of planning determines specific actions. These specific actions are continually measured, reviewed, and updated by all employees in the organization. This activity involves constant attention by leadership and all employees to ensure that the actions are focused and the organization is performing as desired. Action planning becomes part of the everyday method of doing business.

PLANNING

MANAGEMENT

Vision

Mission

Strategy

Tactics

Operations

Typical Planning Flow

EVERYONE

Vision

Action

Action

Mission

Operations

CUSTOMER

Action

Action

Strategy

Tactics

Action

Action Planning Flow

Figure 4.4   Typical planning flow versus action planning flow.

Figure 4.5 TQM planning cycle.

TQM planning cycle

The TQM planning cycle involves an action cycle as shown in Fig. 4.5. During the planning cycle, phase 1 involves defining the focus. In phase 2, the determination of strategic improvement opportunities becomes the task. During phase 3, the selection of critical strategic opportunities for improvement happens. In phase 4, the TQM Plans are developed to maintain a TQM process. These plans provide the specific road map for the organization to strive for victory. During phase 5, results are evaluated. Action planning cycle phase 6 indicates that the complete process be done over and over again in a never-ending cycle. The design action process includes phases 1 through 5 of the action planning process. Phase 6 of the action planning process is the same as the evaluation action process. Chapter 5 details the specific tools and techniques for use during the TQM design process.

Development

Focus: Prepare the TQM system.

Organization: A top-management team—executive team, steering team, core team, or Strategic TQM/Quality Council—consisting of major stakeholders in the organization and lead teams, as appropriate. These teams will need assistance from internal or external resources, as necessary. Internally, the team will need help from various lead teams. In addition, resources are required for coaching top executive(s) and educating, training, and facilitating the various teams. Of particular importance at this point is to develop internal capabilities necessary for implementation. Further, the organization may require support assistance in preparing, conducting, and evaluating various documentation.

Input: TQM plans

Process: Develop TQM system

Output: TQM system

Development action process

The development process involves building the TQM system. The development process performs the actions identified as necessary in the various planning documents for preparing the organization and people for TQM. Specifically, the TQM system requires:

1. Performing project management

2. Documenting the TQM system, as appropriate

3. Establishing the TQM structure

4. Preparing or procuring training materials, as necessary

5. Conducting training of internal trainers, facilitators, and leaders, as needed

6. Performing a pilot project, if applicable

7. Developing "baseline" metrics for strategic objectives

Perform project management activities

Now that the organization has collected information on what the requirements are for its specific TQM system, it needs to manage the project to install TQM. Managing the project involves

- Stating what needs to be done (what)
- Assigning specific responsibility for performance (who)
- Developing a schedule to get it done (when)
- Defining resources
- Calculating risk
- Managing project until completion

Document the TQM system

> *Documentation should support "action" not contribute to "inaction."*

The TQM system must be documented to ensure understanding, conformance, and consistency. However, the TQM system documentation should be the minimum essential for the success of the organization. Chapter 6 describes the processes for the TQM system.

TQM system documentation

TQM system documentation includes the strategic quality plan, TQM system plans, and process plans. In addition, the TQM system might encompass quality system documentation that provides the organization with evidence of internal quality. The quality system documentation should be tailored to the requirements of each specific organization. Quality system documentation could include the following:

- A *quality manual* that gives the overall policy and guidance for the quality system

- *Operation procedures* that outline the general procedures for performing operations

- *Work instructions,* which are the detailed step-by-step methods for performing tasks

- *Records*—the evidence of quality system performance

Documentation action process

1. Document the "as is" process.
2. Determine requirements for documentation.
3. Draft documentation using "as is" process for each requirement for preliminary draft.
4. Identify areas where "as is" process does not meet requirements.
5. Improve the "as is" process to meet or exceed requirements.
6. Update documentation for draft documentation.
7. Perform internal assessment of processes, IAW draft documentation.
8. Take corrective action to change or modify processes.
9. Update documentation into initial final documentation.

Establish the TQM support structure

A TQM support structure is critical in the beginning of TQM. The TQM support structure in some organizations is integrated into its organizational structure. Other organizations, usually the larger organizations or organizations that could not easily integrate TQM into their regular organization, begin with the TQM support organization running alongside the normal organizational structure. Plans should be made up to eventually integrate the TQM support organization into the overall organizational structure or have it form the basis for a new structure. Again, the TQM support organization should be geared to the specific requirements of the organization. Since each organization is different, some organizations will require more support than others.

In order to be effective, the TQM support structure must include the active involvement of all levels of leadership. Top-leadership support is especially critical. The ultimate goal of the support system is to institutionalize the TQM process. It is important to remember that the organization must have support available to accomplish victory.

A support system can have many elements. Typically, a complete support system can include some or all of the elements as shown in Fig. 4.6. These include a coach, owners, steering team, lead teams, teams, mentors, facilitators, and trainers.

The support system should be integrated into the organizational structure. For instance, the steering team would be the executive leadership of the organization. The lead teams would be the middle leaders of the organization or special cross-functional teams formed by the steering team for a specific mission. In addition, teams would be operating everywhere in the organization. These teams could be process improvement teams, quality action teams, work teams, and self-managed teams.

Coach

A coach assists the organization in creating and maintaining the TQM process. The coach is the key support for the steering team and the top executive(s). The coach must be able to "tell it like it is." This requirement, more than any other, usually makes using an outside source for a coach a necessity. Additionally, a coach can help instruct the organization in the fundamentals of TQM, assist in orchestrating the TQM strategy, and function as the TQM overseer for the organization.

A coach is extremely useful, especially in the early stages of TQM, for assisting the organization in transforming the organization from traditional management to a TQM environment. In the initial stages of creating the TQM environment, the coach usually is an outside person to the organization. However, the coach should work with the

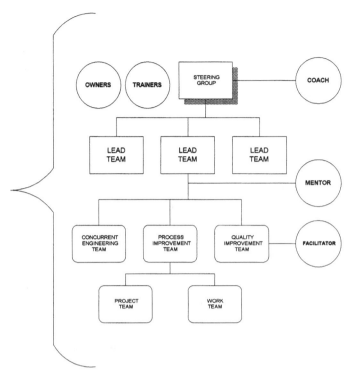

Figure 4.6   Elements of TQM support system.

leaders of the organization to create TQM self-sufficiency in the organization. Remember, the goal is to make TQM an integral part of the organization's way of life.

Owner

The owner is the person who can change the process without further approval. Owners are at every level of the organization. The owner can be part of a team, outside the team, or the team itself. The owner is critical in any improvement effort. The owner must know how the process is performing and continuously look for ways for improvement. In some cases owners can improve the process on their own, or they may need the assistance of others. The owner can frequently receive assistance to perform or improve the process by creating teams.

Steering team

The steering team is an executive-level committee. It is usually composed of the top leaders of the organization. The steering team identifies and prioritizes critical processes for improvement. It provides overall direction, guidance, and support. The steering team also allocates resources as appropriate. It decides on appropriate rewards and recognition. The steering team gives personal attention to all improvement efforts.

Lead team

A lead team oversees several other teams. Lead teams are usually the middle leaders of the organization. Lead teams are normally permanent teams established to perform and improve critical processes. They are the usual major organizational structure for performing the work. Lead teams can be functional or multifunctional teams. In addition, lead teams can create other teams to accomplish a process and/or improve a process. In these cases, the lead team provides leadership for teams improving related processes.

Teams

Teams are groups of people working together toward a common goal. They are an important aspect of Total Quality Management. Teams perform and improve processes. Teams can be permanent like steering groups or the lead teams. Also, teams can be temporary. Typically, many teams are formed in organizations on a temporary basis. These teams are formed to accomplish a specific project or mission. They can also be formed to improve a process. In addition, teams are a normal, problem-solving organizational structure.

Teams can be functional, consisting of all members from the same function, or multifunctional, consisting of members from several functions. They can be composed of any group or combination of groups in the organization, and they offer many advantages over individual performance—for example, they provide better decisions, allow more people to participate, increase communication, and support the systematic, integrated linkage of the organization.

Facilitator

A facilitator assists the team but is not a working member of the team. He or she helps the team concentrate on the mission; ensures that the team stays focused; gives on-the-job training on the use of improvement tools and techniques; provides lessons learned from other team experiences; and assists the leader with team dynamics.

Mentor

A mentor serves as management interface support for a team. He or she functions as a mediator between traditional management and TQM and provides management assistance for a team. The mentor assists the team in obtaining required resources and reviews the team's progress on a regular basis.

Trainers

Trainers provide training as necessary in TQM and job skills. Trainers can also be used as facilitators, often very successfully.

Prepare or procure training materials, as necessary

Training is crucial during the implementation process. Therefore, the training analysis, design, and development must be accomplished during the development process.

Conduct training of internal trainers, facilitators, and leaders

In order to implement TQM, the organization will need the internal capability for training, facilitating, and leading the initial TQM teams. These support personnel should be trained just in time as they are needed to perform their duties.

Perform a pilot project, if applicable

Some small organizations can implement TQM immediately on an organizationwide basis, which is a major change. It requires considerable allocation of resources over a long period of time. For medium- to large-size organizations, performing one or more pilot projects before rolling TQM out for the entire organization is usually recommended.

Some advantages of organizationwide implementation include

- Provides consistent, disciplined approach with common focus
- Fosters the removal of barriers between departments
- Creates a "we" environment at the start
- Shows commitment and support of top management
- Encourages a sense of purpose and urgency
- Demonstrates faith in the process

- Gives some economics of scale
- Develops more in-house capabilities earlier

The advantages to pilot implementation are

- Initial expenditures of resources are less.
- Resistance is minimized by the targeting of locations where champions can ensure success.
- Success builds success.
- Skeptics can be converted by seeing TQM actually work in their own organization.
- Changes can be made based on experience in one's own culture.
- Mistakes can be corrected before TQM is implemented throughout the entire organization.
- The pilot can be used as a model for the remaining organization.

Develop "baseline" metrics

Chapter 8 provides information for developing metrics.

Implementation

Focus: Install TQM system

Organization: The complete TQM structure consisting of steering team, lead teams, process teams, and improvement teams, as appropriate for the organization. In addition, a support structure, including coach, owners, facilitators, and mentors, best fitting the organization should be instituted. The TQM structure will need assistance from internal or external resources, as necessary. Resources are required for coaching, educating, training, facilitating, and supporting.

Input: TQM system

Process: Implement TQM

Output: TQM process

Implementation action process steps

1. Operate as a TQM system.
2. Provide training, facilitating, mentoring, as applicable.
3. Commit resources and support.

4. Form teams, as appropriate.
5. Take action to make improvements.

Operate as a TQM system

Now is the time the organization actually has to operate as a TQM system. This task is not easy. There will be numerous occasions to waiver off the TQM path. Many times TQM will clash with the current organizational culture. There will be many assaults on the TQM system both from within and from outside the organization. Both the organization and the people in the organization must remain steadfast on the TQM course. This endeavor requires top leadership to set an example and peer pressure to be applied throughout the organization. The recognition and rewards system must constantly reinforce the appropriate behaviors and actions. Occasionally, a person or the organization will slip. This error should be recognized, the organization should learn from it, and then it should move on.

Provide training, facilitating, mentoring, as applicable

The TQM plans should include a plan for development of the people.

Commit resources and support

TQM requires additional resources and support. The organization must make the necessary investment in time and money to make the organization a success.

Form teams

Chapter 6 provides details for forming teams and building teamwork.

Take action to make improvements

During implementation, all recommendations for improvement should be given a fair hearing. Although it may not be possible to approve all improvements, the majority of them should be approved; those that are not must have valid justification.

Evaluation

Focus: Check progress of TQM against desired objectives.

Organization: A top management team—executive team, steering team, core team, or Strategic TQM/Quality Council—consisting of major stakeholders in the organization. This team reviews the

TQM progress, periodically. They need the assistance from an internal audit team, as necessary. External assistance for auditing would help this process. In addition, the organization's progress must be assessed continuously.

Input: TQM process

Process: Evaluate TQM process

Output: Continuous improvement

Evaluation Action Process Steps

During evaluation, the organization and teams perform evaluation, assessments, and audits to monitor progress. As a result of the evaluation process, the decision must be made to

1. Continue to perform as is.
2. Use improvement methodology to make it better.
3. Invent a new process.
4. Return to design, development, or implementation process.
5. Start again at assessment.

In addition to the above, the following tasks should be accomplished during the evaluation process:

- Gather lessons learned into an open information system.
- Provide appropriate reward and recognition.
- Maintain relationships.

The evaluation action process consists of the following:

1. Perform continuous assessment of the organization's progress.
2. Assess TQM system.
3. Perform periodic internal audits.
4. Arrange for periodic external audit.
5. Return to the assessment process.

5

Establishing
a Common Focus:
Tools and Techniques

The common focus of the organization encompasses vision, mission, values, goals, quality policy, strategic quality planning, TQM system planning, and process planning.

Establishing and focusing on a common purpose is one of the essential elements of VICTORY-C and a major role of a modern leader. There must be a constancy of purpose throughout the organization. The focus provides the common reason for joint action. When aligned in the organization, the vision, mission, values, goals, and policies provide a common purpose for all to follow.

Establishing Common Focus Considerations

Pursue excellence

Understand what the organization wants to do

Reinforce through action

Provide a common reason for action

Orient toward the customer

Set by leaders, shared by the whole organization

Establish target

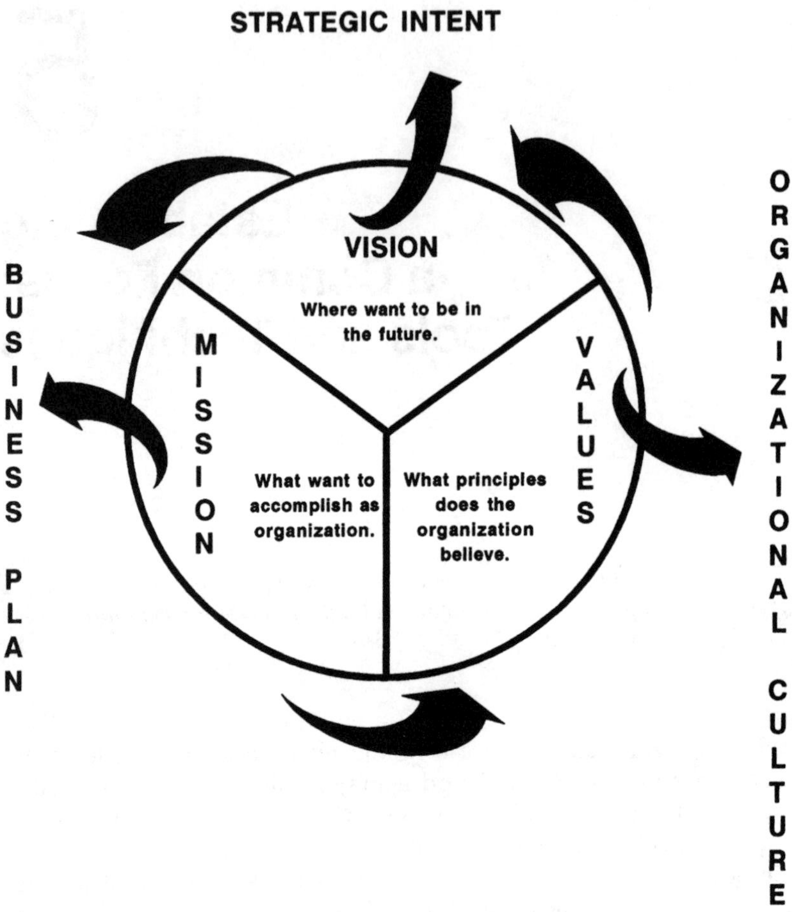

Figure 5.1   Vision, mission, and values.

The focus

The vision, mission, and values form the common purpose or focus of the organization. Figure 5.1 shows the relationship of vision, mission, and values of the organization. The vision is where the organization wants to be in the future. It provides the strategic intent for the organization. The mission states what the organization wants to accomplish; it forms the foundation of the business plan. The values of the organization describe the principle beliefs of the organization and indicate the organizational culture. In addition, goals and policies provide clarification for the focus. Goals also tell how you are doing. Policies provide the guidance.

Creating a Vision

A vision is the organization's view of the future which provides the focus for the journey of the organization. It expresses the ultimate "image" of the organization in the future. It also should reflect the continuous quest for excellence within the organization and constant pursuit to fulfill customer expectations.

Top leadership creates the vision in an organization. Although top leadership usually creates the vision, the whole organization must embrace it. For this acceptance to occur, the vision must have meaning and be shared by everyone in the organization. It must be more than a slogan. It must move the organization to some definite course of action as guided by the organization's leadership through example. To have relevance to where the organization wants to go, the vision has an orientation toward the customer. The vision must communicate both inside and outside the organization a long-term picture for the organization. This vision requires constant communication to build the loyalty and trust necessary to develop a work force committed to its achievement.

Vision considerations

Ｖiew of the future

Ｉnstitute the vision within the organization

Ｓet the example through leadership

Ｉnclude where organization wants to go in the future

Ｏrient toward customers

Ｎurture through constant communication

When to create a vision

A vision is necessary for a new organization, whether it be a company, division, department, project group, or even a team. A shared vision is important when an organization is just starting to provide a sense of purpose for the long-term.

A vision is also helpful in an established organization. It gives a common focus to organizations without a stated future image, or it redirects organizations seeking to adapt to changing environments or the installation of a new leader.

A vision revitalizes an organization that is in a rut or complacent. It does so by providing an opportunity for that organization to look to future successes. That vision sometimes gives the organization the push to go beyond their current situation with some hope and enthusiasm for improvement.

To reiterate, a vision statement may be appropriate when one is

- Forming a new organization
- Starting a project
- Needing to get a clear focus on the future
- Focusing on new priorities for the future
- Adapting to changing environments
- Installing a new leader

A vision is normally not appropriate when one is

- Forming an organization with a short life span
- Creating a committee
- Working on a problem-solving or process improvement team

Vision examples

To be one of the most respected diversified companies in the world.—GenCorp

To have the most effective fighting force.—U.S. Department of Defense

Be the leading supplier to the Products Distribution Base on the North American Continent and the dominant force in every major market in which we compete. To be recognized by the Distribution Base as the technologically innovative leader in logistics support while providing security for all employees and stability for our suppliers and customers.—Elkay Plastics

How to create a vision

The action process steps for creating a vision are as follows:

1. Visualize the future. The top leadership in the organization imagine their ideal view of the future organization. This future view should be years away. Depending on the dynamics of the organization, this time can be anywhere from five years to an unlimited time in the future.

2. Write the vision statement along the top of a flipchart.

3. Review the vision with additional background information. Although the top person in the organization normally provides the initial vision, top leadership in the organization usually further refines the vision to establish commitment and support. Included in this group are the top person, direct reports, and others as deemed appropriate by the organization (e.g., representatives of a union, middle managers, suppliers, and customers).

4. Evaluate the vision. Evaluation is done to ensure that the initial vision meets the organizational criteria of the future. It should answer the following questions:

- Is it an attainable view of the future?
- Does it clearly state to all of the people in the organization a common purpose for the organization?
- Does it convey where the organization really wants to go?
- Does it provide an understanding of what to do?
- Is it oriented toward organizational excellence and specific customer expectations?

5. Clarify the vision statement. Rewrite the vision statement until it is clear, concise, and simple enough for everyone to understand.

6. Institute the vision in the organization. This objective is accomplished through constant communication at every opportunity. In addition, the vision must be set and sustained by the example of everyone in the organization, especially top leadership.

Developing a Mission Statement

The mission describes the basic purpose and expected results of the organization. It sets the common purpose of the organization. For a corporation, the mission describes the corporate view of the role and function of the organization in satisfying customers' expectations today and in the future. For a public agency, mission derives from constitutional or legislative mandates and executive requirements. Mission statements further detail how the organization is going to fulfill its public mandate. For a business concern or public enterprise, the mission is a function of the view of what lies at the core of the organization. Mission is "what we are in business for." For a team, the mission is the intended result of the team's efforts.

Mission considerations

Must be customer-driven

Includes the purpose for the organization

Sets the common direction

Sets the expected results

Involves all stakeholders

Opens and maintains communications

Needs long-term orientation

When to develop a mission statement

A mission should be created for a new organization, whether it be a company, division, department, function, project group, or team. A mission is essential when an organization is just starting to provide a common purpose with clear expectations.

A mission is also helpful in an established organization. It gives a common purpose to organizations or it redirects organizations seeking to adapt to changing environments.

A mission should be created or revised when an organization requires improvement. A mission clarifies the intended results. It provides a chance for an organization to look to specific outcomes. This opportunity gives the organization the impetus to move forward.

Therefore, a mission statement should be considered for an organization when one is

- Forming a new organization
- Starting a project
- Needing to get a clear focus on the specific results
- Focusing on new priorities
- Adapting to changing environments
- Installing a new leader
- Forming an organization with a short life span
- Creating a committee
- Working on a problem-solving or process improvement team

Any group of people working together to achieve some result should have a mission statement.

Mission examples

Organization mission examples:

> *Be a leading international organization supplying just in time quality products, committed to providing total customer satisfaction through service and value, while maintaining a positive work environment and achieving a superior return on investment.*—Elkay Plastics

> *Our mission is to continuously improve the company's value to shareholders, customers, employees and society.*—GenCorp

Team mission examples are

- Provide a deliverable meeting the customer's total satisfaction.
- Continuously improve the deliverable's value to the customer.
- Eliminate errors in order processing.
- Decrease cost of manufacturing.
- Improve assembly workmanship.
- Reduce failure rates of circuit boards.

How to develop a mission statement

The action process steps to developing a mission are as follows:

1. Understand the organization, competition, processes, and customers. The mission is based on the knowledge of the organization. Typically, the people developing the mission have a grasp of the critical strategic factors impacting the organization.

2. Brainstorm purposes and intended results of the organization. First, each team member takes five minutes to write items on two separate pieces of paper: one for purposes and the other for intended results of the organization. Second, the team conducts a round-robin brainstorming session to list items on the flipchart, first for the purposes, and then for the intended results. This task involves each team member, in turn, providing one item from her or his sheet of paper. This step is continued until all items from the individual sheets of paper are listed on the flipchart. Third, items are added to the list during a freewheeling brainstorming session. In this stage the list is opened to everyone to add more items.

3. Clarify ideas. Discuss each item that requires explanation.

4. Agree on items for inclusion in the mission statement. The team needs to decide which purposes should be included in the mission statement:

   - Get agreement on nonnegotiable items.
   - Evaluate each item in terms of yes, no, or maybe as to whether it is a vital item.
   - Evaluate each item in terms of yes, no, or maybe as to whether it is an acceptable item.
   - Resolve vital and acceptable list.

5. Write the initial mission statement along the top of a flipchart. An initial mission statement is usually composed from the nonnegotiable and vital items.

6. Review the initial mission with additional background information. What additional information is required to complete the mission statement? The team needs to ensure that the initial mission statement includes all elements critical to the organization. Also, the list of intended results should be checked to ensure it includes measures of customer satisfaction and organizational performance.

7. Evaluate the mission. The initial mission should be evaluated to ensure it meets the organizational criteria of the future. It should answer the following questions:

   - What are the customer needs or expectations driving the mission?
   - Does the mission clearly state the purpose of the organization?
   - What is the common direction stated in the mission?
   - What are the expected results?
   - Does the mission include purpose for everyone?
   - Was the mission reached by consensus?
   - Is the mission oriented toward long-term viewpoint?

8. Clarify the mission statement. Rewrite the mission statement until it is clear, concise, and simple for everyone to understand.

9. Get each team member's personal commitment to the mission statement. Ask all team members individually: Do they agree with the mission statement and can they support it?

10. Institute the mission in the organization through action planning, follow-up, and constant communication.

Determining the Organizational Values

Values are important, as they guide the conduct of the organization. Values include the principles the organization believes and follows and they are derived from the ethics of the organization. They are the collective concept of what is important and "right" about the organization. Typically, values bring to the surface issues of honesty, trust, and integrity; describe ways of communicating within the organization; guide relationships with the competition, suppliers, and customers; and generally establish ground rules for producing on the promise of the organization. Values have to do with the rights and privileges of management and employees and set the tone for policy and procedure. For instance, if the organization values an internal communication system that designs and develops projects through "concurrent and parallel work," it states this value in its core value document.

Values considerations

Views of what is right in the organization

Appears in people's organizational behaviors

Leads to organizational culture

Uses organizational ethics, honesty, integrity, and trust

Encourages relationships

Supports the quest for quality

Organizational values example

Our commitment to success enhances financial stability, employment security, and reliable support for our customers. We believe the following values contribute to our success:

Desirability. Desirability is enhanced by increasing customer satisfaction through

- Being an industry leader
- Providing quality products and services
- Supplying knowledge and technical assistance

Integrity. We reflect an image in our customer's eyes of honesty, credibility, fairness, trust, and respect. We value our reputation earned by our ethical business practices.

Relationships. We constantly maintain positive business and personal relationships through

- Teamwork
- Communication
- Development of partnerships with our employees, suppliers, and customers
- Commitment to success
- Security for our employees
- Fulfillment of our financial obligations

Profitability. We are committed to optimizing profits through the success of our customers, for the mutual benefit of our stockholders, our employees, and our suppliers.

How to determine organizational values

The action process steps for determining organizational values are as follows:

1. Identify basic values through brainstorming. First, each team member takes 10 minutes to write his or her basic values. Second, the team conducts a round-robin brainstorming session to list items on the flipchart. This effort involves each team member, in turn, providing one item from her or his sheet of paper. This step is continued until all items from the individual sheets of paper are listed on the flipchart. Third, items are added to the list during a freewheeling brainstorming session. In this stage the list is opened to everyone to add more items.

2. Clarify ideas. Discuss each item requiring explanation.

3. Agree on items for inclusion in values statement. The team needs to decide which values are candidates for the values statement:

   - Get agreement on nonnegotiable items.
   - Evaluate each item in terms of yes, no, or maybe as to whether it is a vital item.
   - Evaluate each item in terms of yes, no, or maybe as to whether it is an acceptable item.
   - Resolve vital and acceptable list.

4. Write each value into a values statement. An initial values statement is composed from the list of values. Usually, the values statement includes only the nonnegotiable and vital items from the list of values.

5. Review each values statement with additional background information. What additional information is required to complete the values statement? The team needs to ensure that the initial values statement includes all elements critical to the organization.

6. Evaluate the values statement. The initial values statement should be evaluated to ensure it meets the organizational criteria. It should answer the following questions:

- Does it include all the principles the organization believes and follows?

- Does the values statement clearly bring to the surface issues of honesty, trust, ethics, and integrity?

- Does the values statement foster relationships within the organization and with suppliers, competition, community, and customers?

- Does the values statement indicate appropriate organizational behaviors?

- Does the values statement support the quest for quality?

7. Clarify the values statement. Use focus groups or other means to ensure that the values statement is shared throughout the organization. Rewrite the values statement until it represents the consensus of the organization.

8. Get each stakeholder's personal commitment to the values statement. Ask each team member individually whether he or she agrees with the values statement. Everyone in the organization may have to be included in the questioning.

9. Institute the values in the organization. This effort is accomplished through constant communication at every opportunity and positive reinforcement of behaviors reflecting the values statement.

Setting Goals or Strategic Objectives

A goal is the specific desired outcome(s) of some activity. Goals are important, as they let you know exactly where you are going. Goals provide clear direction and focus. Where vision and mission define the long-term view (i.e., where you want to go in the future), goals establish the short-term look (i.e, each step along the way).

Goals also tell how you are doing, which is critical to helping you stay on-track and make necessary adjustments. Goals help you monitor progress, evaluate situations, and make improvements. Every person in the organization and any group or team should have goals. At the organization's strategic level, goals are the same as strategic objectives.

Goals should be specific, measurable, attainable, results-oriented, and time-bound. They should be clear. Set reasonable goals, but do not set sights too low; they should not be too easy. Nor should they be unattainable. Set challenging goals.

Orient goals toward specific measurable results. Make sure there is specific feedback on the goal outcome. Ensure that goals are linked to organizational objectives or customer requirements. Goals should be set by the people closest to the process, as they are the ones that achieve them. Goals must be accepted by the person or group/team responsible for their accomplishment.

Goals Considerations

Gear to specific results—defined within parameters

Observe by measurement—be able to check outcome

Attain success—challenging, but realistic

Limit to specific time—include time boundaries

Set by process owner(s)—let people closest to process set them

When to set goals

Goals should be set once you understand the complete process, know performance capabilities, and establish customer and organization requirements. In addition to the above, for a group or team, goal setting should be accomplished after the vision, mission, and values statements.

Therefore, setting goals should be considered when one is

- Striving for continuous improvement
- Looking for new objectives after prior goals have been achieved
- Working a project
- Needing to revitalize an individual, group, or team
- Raising the standard
- Trying to establish a purpose
- Fostering accountability and ownership
- Making a change in work performance
- Establishing performance expectations
- Defining milestones
- Entering a new performance period

Goal-setting principles

Provide specific targets; state exact outcome(s) or behaviors.

Recognize and reward results.

Involve the stakeholder or stakeholders in goal setting.

Nurture individual differences for achievement, autonomy, and affiliation.

Commit to the goal.

Integrate goal setting into day-to-day operations.

Pursue challenging goals.

Limit the number of goals to a manageable amount.

Encourage feedback on goals.

Systematize goal setting.

Writing effective goals

The following are examples of effective goals, according to guidelines:

Reduce manufacturing cycle time for assembly A from 6 hours to 2 hours within 1 month.

Decrease errors in quantity required block on order processing sheet from 10 per month to 0 per month in 3 months.

By the end of the year, the ABC company will respond within 24 hours to all customer complaints with a solution satisfactory to company and customer.

The following goal statement is not an effective goal according to the guidelines:

Improve order processing time in customer service.

The statement above is more of a mission statement. It should be rewritten as a goal as follows:

Improve, within 30 days, the time to process an order from order receipt to order ready for delivery—from 3 days to 6 hours.

Strategic objectives

Strategic objectives are goals for the total organization. As such, they should meet all the criteria of any other goal—and then some. An additional consideration for strategic objectives is that they must directly relate to the mission of the organization, as they are essentially a measure for determining its achievement. Therefore, the mission statement is the source for the strategic objectives. The organization needs to determine each element of the mission that it wants to measure. Next, a goal statement or strategic objective statement is written. Then metrics are established for each element.

Strategic objectives examples

Strategic objectives are the critical success factors to achieve the mission. The following are specific examples related to different aspects of the TQM process.

Total customer satisfaction

Customers are totally satisfied with the entire transaction.

Just-in-time supplier

Delivery: Make every delivery "just in time" for each customer.

Inventory: Inventory available to immediately meet 100 percent of critical customer demands and 95 percent of all other demands; the remaining 5 percent must be supplied within 72 hours.

High-quality product

External: 100 percent of deliverables meet or exceed customers' expectations.

Internal: Products are 100 percent defect- and error-free.

Superior return on investment

Meet or exceed 100 percent of financial plan.

Positive work environment

Employees are satisfied with the work environment.

How to set goals or strategic objectives

The action process steps to setting goals are as follows:

1. Understand the mission of the organization, define performance capabilities, establish organization and customer requirements. Before goals are set, it is important to gather critical information affecting the goals. Critical information affecting the goals could be

- Vision
- Mission
- Strategic objectives
- Goals achieved
- Current performance measures
- Customer needs and expectations
- Competition performance

2. Brainstorm expected outcome(s). Do so either as an individual or as a group/team.

3. Clarify ideas. Discuss each item requiring explanation.

4. Agree on items to consider for goals. The individual, group, or team needs to decide which items should be included as goals:

   - Evaluate each item in terms of yes, no, or maybe as to whether it is a goal.
   - Resolve the maybe list.
   - List items for goals.

5. Write an initial effective goal statement along the top of a flip-chart for each goal item. This is the initial goal statement for each goal on the list of goal items.

6. Review each goal statement with additional background information. What additional information is required to complete the goal statement? Were the right people involved in writing the goal? Is there enough information to establish measures?

7. Evaluate the goal statement(s). The initial goal statement should be evaluated to ensure that it meets the criteria for an effective goal. It should answer the following questions:

   - Is the goal geared to a specific result?
   - Can you observe the attainment of the goal by measurement?
   - Is the goal a challenging but reachable one?
   - Is the goal specified to be completed within a certain time period?
   - Was the goal set by the person or people that will make it happen?

8. Clarify the goal statement(s). Rewrite the goal statement until it is clear, concise, and simple for everyone to understand.

9. Get personal commitment to the goal(s). Each person responsible for the goal and/or each group or team member must make the necessary agreement to support the goal(s).

10. Monitor the progress. Establish an action plan to regularly review and take action on the goal(s).

Establishing a Quality Policy

Policies provide guidelines for organizational operations. A quality policy provides the guidelines for organizational quality actions. Each organization requires an overall quality policy. The quality policy ensures a common understanding of the following:

- Organizational meaning of quality
- Quality philosophy
- Quality principles

The quality policy considers following basic quality concepts:

- Quality or customer satisfaction is everyone's responsibility.
- People are not the problem; they are the solution for quality.
- Quantitative methods are the principle means for making decisions.
- A continuous improvement system is the methodology for improving all material services supplied to the organization, all the processes within the organization, and the degree to which the needs of the customer are met, today and in the future.

Quality policy considerations

Provides the guidelines for action

Orients the organization to a quality focus

Launches a quality management system

Inspires a quality organization

Communicates quality throughout the organization

Yields specific outcomes

When to establish a quality policy

A quality policy should be established for a new quality organization, whether it is a company, division, department, project group, or even a team. A quality policy is important when an organization is just starting to provide guidelines for quality efforts.

A quality policy is also helpful in an established organization. It gives a common quality focus to an organization without a stated quality image, or it redirects an organization that is seeking to adapt to changing environments through a quality strategy.

Who should establish a quality policy?

The quality policy should be established by representation for all areas of the organization, including top executives, marketing, manufacturing, logistics, finance, and union representatives.

Determining your meaning of quality

Quality can be defined in many ways, but its meaning often is blurred by the distinction between overall quality, product quality, and service quality. In an organizational context, quality takes on an operational definition as agreed to by the specific organization. In other words, the organization needs to determine its specific definition of quality for the entire organization. This book defines quality as *total satisfaction of customers.* This definition considers quality from a customer's view, and it can apply to any part or the total organization.

What does quality mean in your organization?

The only definition of quality that means anything to an organization is the one that the organization establishes as part of its quality policy. This operational definition of quality should be one that can be applied throughout the total organization.

Deciding on your quality philosophy and principles

Again, each organization must determine its specific quality philosophy and guiding principles to achieve success. Most organizations internalize the most appropriate quality philosophy and guiding principles from one or several of the quality masters. There are many examples of success and failure from using any of the philosophies and guiding principles of the masters. What matters is not so much the one approach, or parts of many approaches, you select as the type of philosophy and guiding principles you select and design for the organization to own and follow to success.

There are many approaches to quality. The major masters of quality are described in Chapter 1 of this book. Each organization needs to research the teaching of these masters. You will notice that there are more similarities than differences among these approaches. As you review the philosophies and principles, the organization needs to determine the specific philosophy and guiding principles items to be included in the organization's quality policy.

FOCUS
Vision
Misson
Values
Quality Policy

STRATEGIC PLANNING
Objectives
Ladders
Goals
Action Items

Figure 5.2   Elements of strategic quality planning.

Strategic Quality Planning

Strategic quality planning involves planning to make the organization's focus a reality. Figure 5.2 graphically shows the elements of strategic quality planning, which involve

1. Create and institutionalize the quality concept that includes:
   - Vision
   - Mission
   - Values
   - Quality policy

2. Perform strategic quality planning:
   - Determine strategic quality objectives. These are the long-term objectives of the organization.
   - Establish quality ladders to meet objectives. This is the five-year plan.
   - Set quality goals. These are the one-year goals.
   - Perform action items. These are the short-term actions achieved in 90 days or less.

Strategic quality objectives

The organization's strategic quality objectives are the goals the organization must achieve to accomplish the mission. Strategic quality objectives include targets for achieving the mission.

Strategic quality objectives action process

1. State the purpose for strategic quality objectives. The focus is to determine the organization's strategic quality objectives.
2. Review mission statement.

Mission statement

Be a *leading* international *organization* supplying *just-in-time quality products,* committed to providing *total customer satisfaction* through service and value, while maintaining a *positive work environment* and *achieving superior return on investment.*

3. List opportunities for strategic objectives.

List of areas for strategic objectives from mission statement

Total customer satisfaction

Leading organization

Quality products

Positive work environment

Superior return on investment

4. Determine metrics to measure each strategic quality objective.

Strategic Objective 1: Total Customer Satisfaction

Desired Outcome: Key customers are totally satisfied with the entire transaction.

Strategic quality objective metric example

Strategic quality objective: Total customer satisfaction

Metric: Customer satisfaction index

Operational definition: On a monthly basis, the strategic quality council will review the customer satisfaction report. The customer service manager will compile the customer satisfaction report.

Measurement method: The measurement consists of a customer satisfaction index compiled as a result of an approved customer satisfaction critique. The customer satisfaction critique indicates the customer satisfaction with:

- Deliverable—supplying what you and the customer agreed upon
- Relationship
- Execution—schedule, cost, technical performance satisfying customer

Desired outcome: Key customers are totally satisfied with the entire transaction.

Linkage to organizational objective: Strategic objective

Measurement owner: Customer service manager

Owner: Strategic quality council

Quality Ladders

Quality ladders are plans for the organization which specify the specific tactics with outcomes to achieve the strategic objectives. A five-year plan is established for each of the strategic quality objectives. Figure 5.3 shows a template for a quality ladder for total customer satisfaction. The five-year plan promotes continuous progress. An example of rungs on the ladder of the five-year plan are as follows:

Year 1   Baseline

Year 2   Improving

**QUALITY LADDER
FOR TOTAL CUSTOMER SATISFACTION**

2000	World Class	
1999	Leader in Industry	
1998	Competitive	
1997	Improvement	
1996	Baseline	

Figure 5.3 Quality ladder template example.

Year 3   Competitive

Year 4   Leader in industry

Year 5   World class

For each of the years above, goals are detailed to meet the strategic objectives. In some organizations, this specification of goals can be expanded to include the goals for each department or process necessary to meet the strategic objectives.

Quality ladders action process

1. Assess current capabilities with baseline metric.

2. Determine tactics for each strategic quality objective needed to get from where we are today to where we want to be in the future.

3. Identify realistic five-year goals using the specific tactics.

4. Document the quality ladders.

5. Perform continuous and periodic reviews.

Quality action items

The quality action items are the specific operational actions that need to be accomplished to achieve goals. These items are actions that are to be accomplished within 90 days to achieve this year's quality ladder or any other improvement action. Action items are updated as necessary; they should be reviewed and updated by management at least every 30 days. An example of a quality action item list is shown in Fig. 5.4.

TQM System Planning

The TQM system is the organizationwide, systematic, integrated, and consistent TQM approach for the specific organization. TQM systems planning involves

- Identifying the elements of a TQM system for your specific organization

- Establishing responsibility for each of the TQM system elements

- Determining a schedule for accomplishing each of the elements

Identifying specific TQM system elements

The first step in developing a TQM system is to define the specific requirements of the system for the particular organization. This builds

Item #	Action Item	Date to Start	Date Completed	Owner	Status or comments

Figure 5.4 Action item record.

on the Strategic Quality Plan to identify the TQM system elements. The specific TQM system should always consider the particular organization's focus; customers' needs and expectations; and risks, costs, and benefits.

There are many sources of information for a TQM system. A simple TQM system would address each of the elements of VICTORY-C, which would be sufficient for many organizations. However, many industries, like medical, defense, and manufacturing, have established standards for quality. These may be the minimum requirements for being recognized as "qualified" to do business in a particular area. Therefore, they must be considered as a source for elements of the TQM system. For instance, the ISO-9000 series of standards or ANSI/ASQC Q9000 series for a quality system provide a good foundation for a basic quality system for consistent "product or service" quality. ISO-9004/ANSI/ASQC Q9004-94 specifically details guidance for this type of quality system. This is a good starting point for identifying elements for a TQM system for organizations that need a basic quality system, require compliance to the standard for business purposes, or want to be consistent with a common standard. Remember, the ISO-9000 Standard series focuses mainly on product quality. Continuous improvement is only a secondary goal. A TQM system needs to go beyond mere product quality to total customer satisfaction. This effort requires balancing consistency with flexibility, which can be achieved by building controls where necessary for product quality while developing the people to become empowered to make continuous improvements within the system. Therefore, you may need to expand upon the elements of the ISO-9000 series by going beyond the basics in each element and adding additional elements, as appropriate, to build your own TQM system.

Other sources of TQM elements include award criteria like the Malcolm Baldrige National Quality Award. This award criteria is widely recognized by the nation as containing the major elements for quality. In fact, this award often equates to TQM in many people's mind. It is a more complex version of the VICTORY-C model.

In general, the organization should review the following elements for inclusion in a TQM system:

- VICTORY-C elements:

  Customer focus and total customer satisfaction

  Modern leadership

  Vision

  Involvement of everyone and everything

  Continuous improvement

  Training, educating, facilitating, coaching, mentoring

  Ownership

  Recognition and reward

  Yearning for success

- ISO Standards/ANSI/ASQC Standards:

  Management responsibility

  Quality system

  Contract review (quality in marketing)

  Design control (quality in engineering)

  Document and data control (quality of document control)

  Purchasing (quality in procurement)

  Control of customer-supplied product

  Product identification and traceability

  Process control (quality in production)

Inspection and testing (product verification)

Control of inspection, measuring, and test equipment

Inspection and test status

Control of nonconforming product

Corrective and preventive action

Handling, storage, packaging, preservation, and delivery

Control of quality records

Internal quality records

Internal quality audits

Training (personnel)

Servicing

Statistical techniques (use of statistical methods)

Quality-related cost considerations

- Malcolm Baldrige National Quality Award Criteria:

  Leadership

  Information and analysis

  Strategic Quality Planning

  Human resource development and management

  Management of process quality

  Customer focus and satisfaction

  Quality and operational results

- Industry standards
- Organizational structure with support
- Standards for workmanship
- Supplier and customer relationships
- Customer service
- Quality control for product and/or service quality
- Quality assurance for product and/or service quality
- Product support
- Safety
- Metrics and measurements
- Information systems

Again, the TQM system should contain the elements necessary for specific organizational success. The focus should always be on the mission of the organization. The TQM system is only the means to the organizational ends. If there is not an organizational reason, it should not be an element of TQM.

Develop a TQM system schedule

Once the elements are identified for the TQM system, the tasks, owners, and schedule must be established to ensure the correct management of the project. This effort involves first identifying the current state of the organization (see outline in Chapter 4), then establishing the tasks to be accomplished for the organization to go from where it is today to where it wants to be in the future.

Process planning

Process planning constitutes the everyday planning activities for getting the work done in an organization. The purpose of process planning is to record the important elements of a process to explain how it is currently performed within the organization.

Process planning action process

The process plan action process contains the following:

1. *Process description.* This statement shows the overall process in terms of the customer (internal/external) and internal operations. First, the process is stated as an action and a descriptor. For instance, the customer service process might be stated as "serve customers." Second, the general operation performed in the process is stated.

   Example of Process Description: Serve customers by providing quotes, processing orders, and keeping them happy.

2. *Process Start.* The action that begins the process.

3. *Process End.* The action that ends the process.

4. *Process operations.* All operations necessary to perform the process.

5. *Supplier requirements.* The input to the process and who supplies the input. Requirements should be as detailed as necessary.

6. *Customer expectations.* The output of the process. These outputs are expressed as products or services to customers (internal/external). State as specifically as possible the customers' needs and expectations from the process.

7. *Process measure(s)*. The process measurement or metrics. They help us to know if the process is performing up to the organization's standards while meeting customers' expectations.

8. *Responsibilities*. The general responsibilities needed to perform the process are listed.

9. *Resources*. The major resources (people, equipment, materials, etc.) needed to perform the process are stated.

10. *Education and training requirements*. The specific education and training qualification to perform the process are listed.

6

Involving People:
Tools and Techniques

The organization that maximizes its human resources wins.

This chapter contains the following tools and techniques:

- Individual involvement
- Teams and teamwork
- Managing conflict
- Forming a team
- Meetings
- Brainstorming
- Consensus decision making
- Presentation

Introduction

People are the key to success in Total Quality Management. They perform and improve the processes. Total Quality Management aims to
maximize the potential of human resources in an organization, specifically by fostering both individual and team contributions to the organization. It relies on individuals working smart and taking pride in
their work, accomplishing the mission. In addition, these individuals'
contributions are multiplied through teams. People involvement tools
and techniques include individual involvement, teamwork, communication (especially listening), focus setting, meetings, brainstorming,
and presentation.

Individual Involvement

Individual involvement concerns each person's contributions to the organization. In Total Quality Management, individuals work to continually perform their work and improve the processes in the organization focusing on total customer satisfaction.

TQM seeks to benefit from each individual in the work force. All individuals are different, unique, and valuable. In today's economic environment this diversity is a distinct advantage to the organization that learns to use it to improve their competitive position. People have a variety of attitudes, beliefs, perceptions, behaviors, opinions, and ideas. These are potential sources of creativity. Innovation can be gained from the work force's different competencies, abilities, knowledge, and skills. Each person's culture, background, and personality fosters an individuality that can be used for the good of the organization. Creativity, innovation, and individuality can be the edge needed for growth. Therefore, in a TQM organization, individual differences are valued as an important resource.

Although each person is different, people generally want some of the same basic things. They want to be safe and secure, feel trusted, belong, be appreciated, feel important, have pride in work, be involved, and have advancement and personal growth opportunities. The organization that provides a work environment where all of these wants can be achieved by the individual will be rewarded with high individual productivity.

In Total Quality Management, the goal is the actual ownership by everyone in the organization. Ownership means that all individuals in the organization do what is necessary to perform and improve their work. Ownership does not just happen. The organization cannot simply announce it and expect it to work. Typically, ownership comes in stages. First, people must trust the organization. Typically, most organizations have developed many adversarial relationships over the years, which has led to mistrust between management and workers, organizations and unions, and one department or function and another. This barrier must be removed before an individual will become involved in any extraordinary effort. Restoring trust may take some time, depending on the organization. This goal can be achieved only by the actions of management working through structured activities. These activities should foster honest and open communication leading to some specific actions that build the trust.

Once trust is restored, people will begin to become involved in assuming more ownership of their work. At this point, resources must be available to allow the people to take pride in their work. When pride in the work is the norm, people take the ownership to provide total customer satisfaction.

With added emphasis on human resources, people must work smarter to perform and improve their work with a focus on customer satisfaction. People have always known best how to do things right— and do them better. However, historically neither the organization nor the people have known how to tap this resource for the benefit of the organization, the individual, and the customer. The organization must be transformed to provide an environment where individuals can maximize their potential. At the same time people must be trained in a systematic process that provides them the capability to influence their work. When this goal is accomplished, individual involvement can reach its maximum potential.

Individual involvement is fostered by the following:

Install pride of workmanship

Nurture individual self-esteem

Develop an atmosphere of trust and encouragement

Involve everyone

Visualize a common purpose

Improve everything

Demand effective and open communications

Use recognition and rewards

Allow creativity and innovation

Lead by example

Individual involvement action process

1. Establish a people-centered environment
2. Provide development opportunities
3. Provide experiences with expected behavior
4. Recognize and reward appropriate behavior, actions, and results

Teams

A team is a group of people working together for a common goal. Teams should not be confused with groups. A team shares responsibility, authority, and resources to achieve their collective mission. They feel empowered to do whatever is necessary within their defined boundaries. Action through cooperation is practiced both within the team and when acquiring support. Problem solving and decision making are natural activities practiced by the team. Effective, open, and full communication, especially listening, is prolific. The leader and the members possess a positive "can do" attitude even during difficult times. Team members motivate, respect, and support one another; manage conflict; build self-esteem and motivate other team members; contribute technical competence in their specialty as well as other skills; and acquire many skills to accomplish the mission and build and maintain teamwork. Effective teams realize that diversity, individuality, and creativity are their greatest advantages. Individual and team contributions are recognized and rewarded appropriately. The team takes ownership and pride in their performance. Everyone is totally committed to cost, schedule, and quality standards of excellence with total customer satisfaction the primary focus of all team activities.

Types of teams

Teams can be either functional or multifunctional. A functional team consists of members from the same discipline or organization. For example, an engineering functional team would be a team in which all the members work in the engineering department. A multifunctional team would have members from engineering, manufacturing, marketing, and other departments.

Teamwork

Teamwork is the technique in which the individual team members work together to achieve a common goal. This process involves cooperative relationships, open communications, group problem solving, and consensus decision making. Teamwork can only be effective in an environment of honesty, trust, open communications, individual involvement, pride of workmanship, and commitment. Specifically, effective teamwork involves the following:

Trust

Effective communication, especially listening

Attitude positive—"can do"

Motivation to perform and improve

We mentality

Ownership of work with pride

Respect and consideration of others

Keeping focused on total customer satisfaction

Benefits of teamwork

Teamwork provides the responsive work force required to survive in today's environment. Cooperation toward a common goal is essential for success. Some of the benefits of teamwork include:

Better decisions and motivation

Everyone can participate

Nurtures improved working relationships

Encourages rewards in work itself

Freer contribution of information

Increased communication

Thrusts an organization toward common focus

Supports an organizationwide perspective

Principles of teamwork

In order to build and maintain teamwork, the team must obey some principles. The key principles of teamwork involve the following:

Keep focused on the mission, not on the person

Encourage open communication and active listening

Yearn for constructive relationships

In addition to these key principles, there are basic principles that every team must observe to build and maintain teamwork over the long-term. The team must be continuously developing and maintaining teamwork. The individual team members and the team must receive appropriate recognition and rewards to maintain interest in teamwork. Further, all members must be involved in team activities to maximize the true potential of the team. Team members must have enough self-esteem to actively contribute. Communication is essential in any team activity. In addition, the strength of the team lies in the individuality of each of the team members. Constructive cooperative relationships are critical, both within and outside the team. Relationships are important between team members and with customers, suppliers, and other teams. All the members, especially the team leader, must set the example. Team members can develop the behaviors necessary to work as a team through observation. Ideas are the power of the team. All team members must be encouraged to continually contribute innovative and creative ideas. Above all, the focus must be on the mission, not the person. Teamwork is not personal; instead, it demands a unrelenting devotion to a common purpose. The basic principles of teamwork can be summarized as follows:

Pursue team environment

Recognize and reward the individual and the team

Involve all team members

Nurture the self-esteem of all team members

Communicate freely and openly

Include individuality

Pursue constructive relationships

Lead by example

Encourage ideas from all team members

Stay focused on the mission

Building teamwork

Team building requires a continual diagnosis and improvement of the effectiveness of the team. In order to build the cohesiveness and effectiveness of the team, attention must be paid to the mission, roles and responsibilities, group dynamics, and interpersonal relationships within the team.

The following are essential to build teamwork:

- Identify the team mission.
- Establish team roles and responsibilities.
- Understand team dynamics.
- Manage conflict.
- Provide motivation.
- Build individual self-esteem.
- Develop the team.

Identify the team mission

The mission is the intended result. It provides the focus for all team activities, gives the expected outcome(s) of the project, and provides an indication of the magnitude of the project. It should state the boundaries of the project to include specific process(es). More importantly it should define the authority of the team. The team's resources to accomplish the mission must be identified as well. Normally the mission originates from outside the team. It comes in general terms from a variety of sources, for example, from management or customers. The general mission must be negotiated and clarified by the team.

The first outcome-related activity of the team should be clarification of the mission, as it provides the common purpose. The mission must be written in a mission statement which is understood, clear, and achievable. The team must reach consensus on a mission statement before doing any other team activity.

Establish roles and responsibilities

Roles and responsibilities are the specific contributions expected from each team member to accomplish the mission. These contributions can include any formal or informal offerings each team member is to bring to the team. Formal contributions include the expected roles and responsibilities of a specific discipline, function, or organization. Informal offerings are the contributions a team member can add as a result of personal strengths. Each team must develop its own unique roles and responsibilities based on the requirements of the mission and the capabilities of the team members.

Individual team members must have distinct roles and responsibilities with corresponding accountability; these may change as the team develops and the project progresses. The roles and responsibilities must be defined in a "living document" developed by the team, which should be the next team activity after the mission statement has been agreed to. The roles and responsibilities should include:

Results expected—outcome(s) from each team member

Ownership—including the amount of control

Limits of resources—funds, equipment, and people

Empowerment with amount of authority

Standards focusing on customer satisfaction

The roles and responsibilities should include the expected outcomes from each team member. They should be stated in terms relating to the contribution to the mission, if possible, in terms of metrics. In the initial stage of a project it may not be possible to include specific measurement, but performance measurements must be included as soon as possible. They allow individual team members to know exactly what they need to do.

Another part of roles and responsibilities involves ownership. The roles and responsibilities must state which processes each team member owns. This statement provides individual team members with details of what they do.

Critical to performance of roles and responsibilities is the amount of resources available. Again, they should be stated. This statement provides each team member with details of what is available to help achieve the mission.

Empowerment involves having the responsibility, authority, and resources to do whatever is required to satisfy the customer and achieve the mission within defined boundaries. The key to empowerment is defined boundaries; each team member must be familiar with them, as they provide him or her with a statement of what team members can do. These boundaries will change as the team develops and the project progresses. In the beginning of a project, team members usually do not have the capability to be fully empowered. As they are trained and gain new experiences, the team can assume more empowerment. Eventually the team can be fully empowered. This stage is when the maximum potential of the team can be realized through the creativity and innovation of the team members.

Standards are an essential part of roles and responsibilities. They are the accepted norms for all team members that focus on customer satisfaction. Standards must be a clear definition of what is acceptable under all situations. This statement provides individual team members with details of what they all should do.

Specific team roles and responsibilities

The team consists of a team leader, team members, and sometimes a team facilitator. Each of these team players has a specific role. The team leader guides the team to mission accomplishment. The team members contribute toward achieving the mission. The team facilitator assists the team with focus, teamwork, methodology, tools, and techniques.

The type of team leader and team member roles depend on the category of the team. Figure 6.1 shows four categories of teams. The first category of team is the traditional directive organization with a manager. With this team the role of the manager is to get the task accomplished and the role of the team member is strictly to perform the directed job. The second category of team is a participative organization. With this type of team, a leader guides the team to a common goal through a process involving all team members; the team members provide their expertise and cooperation. The third category of team is a collective self-led organization. In this team, ownership is shared by all team members; a team facilitator creates and maintains teamwork. The fourth category is an empowered organization. In an

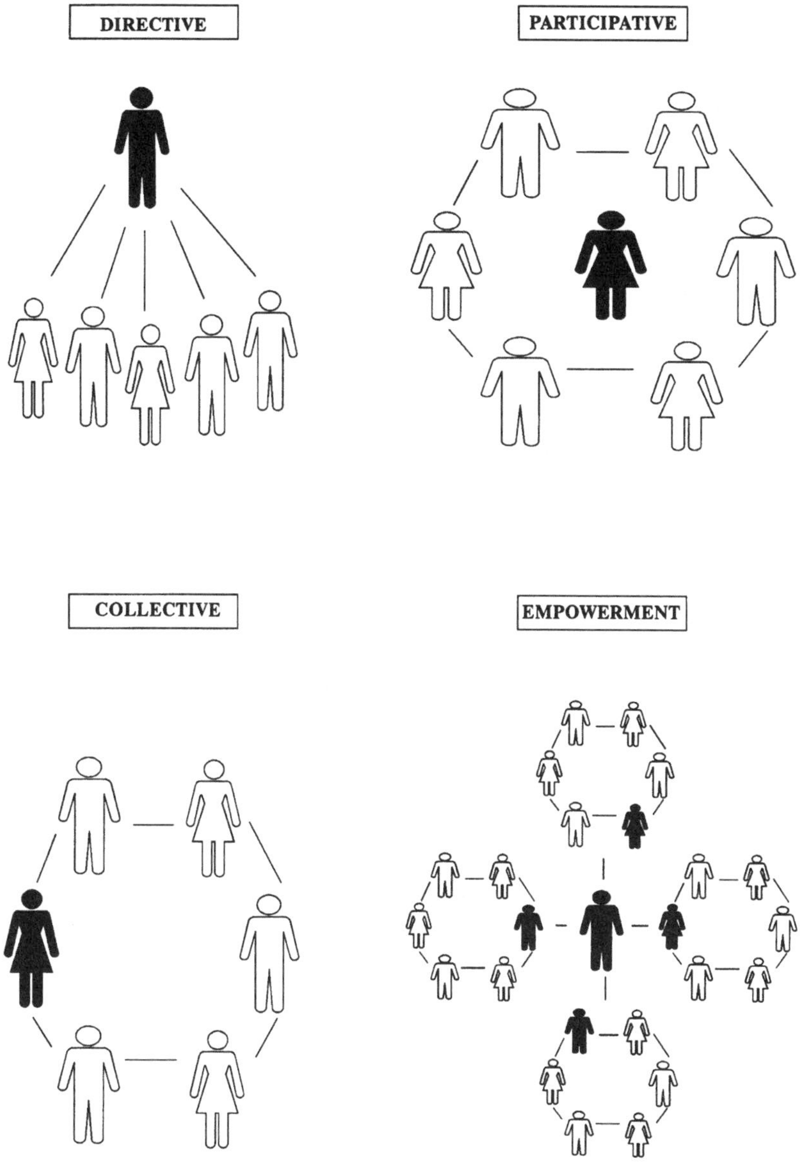

Figure 6.1 Four team categories.

empowered organization, teams have the total responsibility, authority, and resources to perform and improve their process(es). In this category of team organization, a coach, mentor, and/or resource person advises the team.

Understand team dynamics

Each team must understand that although it is unique, all teams normally go through four distinct stages before they are truly performing as a team. As shown in Fig. 6.2, the four stages of team development are orientation, dissatisfaction, resolution, and production.

Each team must go through all four of the stages of team development before it reaches synergy. There is no shortcut. The duration and intensity of each stage varies by team. It is important to maintain the focus and a positive attitude throughout all the stages; the

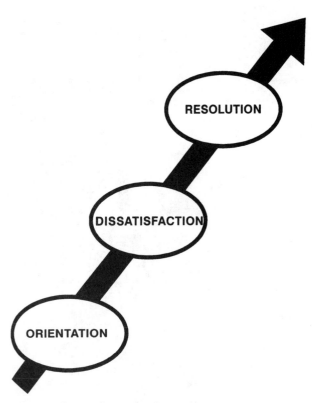

Figure 6.2 Stages of team development.

team will achieve its mission. Below is a general description of each of the stages of team development.

Stage 1: Orientation. During the first stage, team members are becoming acquainted with one another and teamwork. Members are building rapport, honesty, trust, and open communication. They are trying to determine what it takes to fit in. The team members usually have great enthusiasm for the project. However, they do not know how to work as a team to accomplish it. During this stage the team is deciding what they need to accomplish and who needs to accomplish it.

Stage 2: Dissatisfaction. Stage 2 is characterized by the team being overwhelmed by the information and task. Sometimes power struggles, emotions, and egos become evident. This stage is the most difficult to overcome. Some teams never progress past this stage. If they don't, they should be disbanded. To move forward to the next stage, the team must find some small success as a group. Once the team understands that they can perform as a team, they usually progress to the next stage.

Stage 3: Resolution. During stage 3, the team moves toward the mission. In this stage customer contact and measurements can help team members start to assist each other and focus on the mission. This is the first stage in which the team is actually working as a team. At this time the team knows how to operate as a team.

Stage 4: Production. Finally in stage 4 the team becomes effective. The team members work together to achieve the mission.

Manage conflict

Conflict can exist whenever two or more people get together. Differences exist in every organization. These differences are an advantage to any organization that has learned to control conflict. Conflict may be positive and agreement negative.

Differences exist in every organization. Our diversity is one of our major strengths. In teams, we must take advantage of differences to be successful. For example, an organization can select from its diverse work force a group of workers from a specific culture and background and use them to research a potential new market or product targeted toward people of that particular makeup. Further, the organization can gain new ideas from a diverse work force, which can lead to improved operations, decreased cost, and/or reduced time. The following is a list of potential sources of conflict that can be beneficial to an organization:

Cultures and background

Opinions

Needs and expectations

Facts and perceptions

Levels, departments, and organizations

Interests, personalities, and egos

Competencies, knowledge, and skills

Targets, missions, goals, and objectives

Conflict can be controlled in the following ways:

Cooperate rather than compete.

Orient toward the issue; not the person.

Negotiate win/win solutions.

Take an organizationwide perspective.

Recognize conflict as natural.

Observe empathy with other's views.

Limit perceived status differences.

Conflict can be positive in many ways. With conflict, the team does the following.

Pursues win-win situations

Observes others' points of view

Shows open communication

Instills an organizationwide view

Takes personalities out of the issue

Invites trust and involvement

Views entire issue

Examines different sides of an issue

All this leads to effective consensus decision making, which establishes and maintains teamwork.

Agreement can be negative, especially when it involves what is commonly called groupthink. Groupthink is the tendency of the group to agree and may have an adverse effect on the effectiveness of the team to achieve the mission. Groupthink comes from many sources. Sometimes groupthink results from the good intention of maintaining the cohesiveness of the team. In other cases groupthink stems from fear. The team members may be afraid of losing their jobs, losing face, or offending the leader, management, or other team members. Regardless of the source, groupthink must be identified and controlled. The following are some specific actions to overcome groupthink:

Appoint a devil's advocate.

Get open discussion on all issues.

Recognize impact of status differences.

Examine all agreement without resistance.

Evaluate all views of the issue.

Conflict symptoms

The first step in managing conflict is recognizing that conflict exists. Everyone on the team, including the team leader, team members, and especially the team facilitator, must be constantly alert to the symptoms of conflict and groupthink. Some of the symptoms of conflict and groupthink are

Stopping open communication

Yielding to win-lose solutions

Making little movement toward solution

Pressure to stop challenges

Taking sides (we/they)

Observing no building on suggestions

Members silent

Stopping any resistance

Conflict management actions

Conflict can be managed during the day-to-day operations of the team. First, avoid any face-saving situation. If honor and pride are at stake, people will defend their position even when they realize that they may not have the answer. Second, continuously self-examine attitudes. Sometimes a person may develop an attitude triggered by an emotional response. This situation may be detrimental to teamwork. Focus on the mission and maintain a positive attitude throughout all team activities. Third, target win-win solutions. This effort allows the team to avoid we/they situations. Fourth, involve everyone in all team activities. People do not always agree with one another's contributions. But if all team members participate, they will support the decision. Fifth, observe the limits of arguing: it is useless and does not lead to positive solutions. Sixth, nurture differences of opinion. Everyone is right in her or his own mind. There are no right or wrong answers. Differences of opinion can be used to stimulate other ideas. Seventh, support constructive relationships. Relationships are the key to all teamwork. Build long-term relationships on a foundation of honesty and trust. This endeavor will allow open and free communication which is the real key to conflict management.

Conflict management actions are as follows:

Avoid face-saving situations.

Continuously self-examine attitudes.

Target win-win solutions.

Involve everyone.

Observe the limits of arguing.

Nurture differences of opinion.

Support constructive relationships.

Provide motivation

Motivation is the behavior of an individual whose energy is selectively directed toward a goal. Performance is the result of having both the ability and the motivation to do the task. Motivation influences team members to certain behavior. Motivation depends on satisfying the needs of the individuals. Traditionally, motivation equated to extrinsic rewards such as compensation, promotion, and benefits. These were aimed at satisfying the basic needs of individuals for housing, food, and clothing. Today, people need to be motivated by a higher order of needs: intrinsic needs such as a sense of belonging, a feeling of accomplishment, improved self-esteem, and opportunities for personal growth. Teamwork, especially on customer-driven teams, provides rewards for satisfying intrinsic needs.

Recognition and rewards for individual and team performance is essential for teamwork. Lower-level rewards are usually sufficient motivators when teams are started. Once a team is established, team members covet higher-level intrinsic rewards—for example, personal development workshops. During all the stages of team development, recognition is particularly effective to reinforce positive behaviors. Praise and celebrations are necessary to maintain teamwork. Some examples of recognition include letters of appreciation, a pizza party, coffee and donuts, or a public announcement. Particularly effective is a pat on the back with a comment such as "you did a good job." In the early stages, extrinsic rewards have a short-term effect; however, they may actually be a negative motivator for long-term teamwork. Extrinsic rewards are often important for long-term teamwork, but they must be appropriate for the desired outcomes. Before any rewards are instituted, they must be thoroughly analyzed to ensure fairness to everyone.

Besides recognition and rewards, motivation can be provided by the team to team members. The following are some specific actions the team can use to motivate team members:

Make it clear that the goal is shared.

Orient, develop, and integrate team members.

Think and speak "we."

Institute internal team rewards and recognition.

Value individual contributions.

Avoid frequent changes of team members.

Take time to develop relationships.

Encourage sense of belonging.

Build individual self-esteem

An individual's self-esteem affects his performance of organizational tasks as well as his relationship with others on the team. There are actions that each team member can do to maintain and build the self-esteem of team members. They are as follows:

Establish an environment in which an individual feels that her self-worth is important to performance.

Stay focused on the mission; do not make it personal.

Treat each person as you want to be treated.

Encourage individual contributions.

Ensure individual recognition and reward.

Motivate, communicate, involve, and develop.

Teamwork critique

Periodically the team should perform a self-assessment of their development as a team. Each team should develop their own critique based upon their criteria of a successful team. This critique should be completed individually, the results tabulated, and evaluated and discussed as a team. The teamwork critique should be performed on a regular schedule. Figure 6.3 shows an example of a teamwork critique.

Forming a team

Forming the team is an important first step—and vital to making teams work. It sets the stage for all other team activities. During this critical phase the team develops the foundation for working together toward a common goal. In many teams, success or failure is determined at this time. When a team is being formed, there should be consideration for both relationship building and performance. However, it is necessary to spend more time in the beginning on relationship building. Resist the temptation to jump right into performing the task. An investment in relationship building will pay huge dividends later in the process.

Forming the team involves

Establishing a shared purpose

Conducting effective meetings

Building teamwork

Deciding on a common process or methodology

For many teams, education and training are also essential for development of team-building knowledge and skills during the formation stage. Specific training should be conducted at the time it is needed for it to be efficient and effective. It is wise to integrate the training requirements into the team process.

Focus on a shared purpose.

Orient team members to one another and to the team process.

Recognize the expected outcomes.

Model a common process or methodology.

When to form a team

Teams are the organizational structure of choice for flexible, rapid response to ever-changing customer needs and expectations. Therefore,

Teamwork Critique

Instructions	Please rate the team based on the 5 point scales below. Circle the number on each scale that best states your opinion at this time. Discuss with team.

1. Trust

Is the level of trust among team members sufficient to allow open and honest communication without tension?

close/ tense	1	2	3	4	5	open/ relaxed

2. Effective Communication, especially listening

Do team members listen to each other?

members do not listen	1	2	3	4	5	members listen

Does everyone have a chance to express their ideas?

no ideas expressed	1	2	3	4	5	variety of ideas expressed

3. Attitude, positive "can do"

Do team members display a willingness to take risks?

avoid risk	1	2	3	4	5	take risk

4. Motivation

Are team members actively participating?

bored/ withdrawn	1	2	3	4	5	involved/ interested

5. "We" Mentality

Do team members demonstrate a togetherness in words and actions? Are decisions based on consensus?

individual contribution/ no consensus sought	1	2	3	4	5	team action/ consensus

6. Ownership

Do team members take the initiative to solve problems and/or improve their process as a natural course of action?

only do what told	1	2	3	4	5	take action to make things better

7. Respect, consideration of others

Do team members respect differences? Are people's differences managed to the team's advantage?

avoid others/ conflict	1	2	3	4	5	respect others/ manage conflict

8. Keeping Focused

Does the team remain targeted on vision, mission, and goals?

off target	1	2	3	4	5	on target

Figure 6.3 Teamwork critique.

a team should be considered when one is

Starting a project

Solving a problem

Performing a mission

Setting strategy

Implementing a new program

Needing high-performance, flexible, adaptable structure

Involving many different people to achieve a common goal

Empowering a work group

Supporting a systematic integration of the organization

A team is normally not appropriate when one is

Reacting to a crisis

Seeking to complete a short-term objective

Working on a project that would be better handled on an individual basis

Trying for excessive control of process and people

For some activities individual contributions, work groups, committees, or task forces may be more appropriate to use.

Preparing to form a team

Preparing to form a team involves

Establishing the purpose of the team

Determining why a team is the most appropriate structure

Deciding who should be part of the team

Establishing the purpose of the team

The purpose is the initial mission statement of the team. It should be drafted before the team is formed. The purpose should

Provide the project scope

Unite the team

Recognize the desired outcomes

Prevent misunderstandings of what is to be accomplished

Orient the team toward specific customer expectations

Set the common direction

Empower—authority, responsibility, and resources

Determining why a team is the most appropriate structure

Use of a team may not always be necessary. Although teams are the organizational structure of choice, it may be more appropriate to take individual action, establish a task force, or use a committee. Teams should be formed when they would

Reinforce business and individual objectives

Encourage participation

Anticipate objections

Solicit commitment to achieve the mission

Obtain support for the team process

Nurture both relationships and results

Deciding who should be part of the team

The team leader is critical to the success of a team. Since the team leader guides the team to work together to achieve a common purpose, she or he should be selected first. Then, the team leader should determine who should be team members. The specific composition of the team depends on the team's mission. The team should include representatives of all stakeholders in the process, for example, customers, suppliers, process owners, process workers, and union leaders. In addition, the number of people on the team should be kept to the absolute minimum needed to achieve the mission.

Selecting the team

The success of the team depends upon the people on the team. The team leader and team members must work together to ensure that the team achieves its mission. Selection criteria for the team leader and team members varies by team. It is important to decide on specific criteria for the team leader and team members for each team.

In many organizations, you will not be able to find the "ideal" candidates. The criteria listed below are the basic characteristics desired in players in the game. In addition, team leaders and members also need to possess or acquire many team skills to be competent. Therefore, the team, team leader, and team members usually require additional assistance. The team should be provided help in coaching, education and training, facilitating, and mentoring as necessary. Further, it is a good idea to systematize an ongoing team leader and team member development process in the organization.

Team leader selection criteria example

Communicates, allows input, is willing to listen

Interested, supportive, appreciative, humanistic, considerate

Displays trustworthiness, honesty, integrity, ethics

Is objective, open-minded, tolerant, reasonable, fair

Delegates, trusts, empowers, allows room to achieve

Motivates, challenges, inspires, is team oriented

Team member selection criteria example

Positive attitude

Willingness to participate

Flexibility and adaptability

Ability to do and follow at same time

Stakeholder, owner, or expert in the process

Team kick-off meeting

The initial or "kick-off" meeting is critical to the future success of the team. First impressions last for a long time. The team leader must take the time to have an excellent kick-off meeting. Figure 6.4 provides a kick-off meeting checklist.

Forming a team action process

1. Provide the purpose of the team.
2. Link the purpose of the team to business focus.
3. Provide an opportunity for members to discover "what's in it for me."
4. Take time to get to know one another.
5. Learn how to conduct effective team meetings.
6. Prepare a team code of conduct.

Team's initial mission statement	☐
Reason for the team	☐
Team leader selection	☐
Team members selection	☐
Kick-off meeting set-up	☐
Determine meeting date	☐
Schedule meeting room	☐
Arrange for introduction by management	☐
Letter of Invitation	☐
Meeting Focus Statement (attach to Invitation)	☐
Meeting Agenda (attach to Invitation)	☐
Meeting room set-up (chairs, overhead, easel, etc.)	☐
Meeting materials (handouts, chalk, markers, etc.)	☐

Figure 6.4 Kick-off meeting checklist.

7. Determine team meeting roles.

8. Establish a process for conducting meetings.

9. Learn the fundamentals of teamwork.

10. Clarify or write the team's mission statement.

11. Define the roles of team members necessary to achieve the mission.

12. Establish a methodology for accomplishing the mission.

Meetings

Meetings are a technique of bringing a team together to work for a common goal. Effective meetings are an important aspect of Total Quality Management in that they get the team to develop improvements that an individual could not come up with. By bringing people together in a meeting to develop improvements for a common goal, better decisions can result. The key is to make the meeting effective. Effective meetings require an action-oriented focus. All the members of the team must have a common focus and a common methodology geared toward specific actions.

Meetings can be effective through the use of meeting tools. The meeting tools provide rules of conduct; identify meeting roles, responsibilities, and relationships; give a focus; and provide documentation of progress. The meeting tools are

- Rules or code of conduct
- Meeting roles, responsibilities, and relationships
- Focus statement
- Agenda

Meeting considerations

Make a focus statement.

Ensure that team meeting roles are assigned and understood.

Ensure that the team uses an agenda.

Take time to prepare, participate, and perform.

Rules of conduct

Rules of conduct provide guidance for the team's conduct. The code of conduct considers "how" meetings will be conducted. Each team makes its own unique rules of conduct. These rules are determined during the first team meeting by consensus. The rules of conduct open communication for the team in a nonthreatening situation. They are posted during every team activity. Although they are established during the first team meeting, these rules can be changed at any time the team deems necessary. However, the rules are established by consensus in the first meeting to help build rapport in a nonthreatening task.

Code of conduct

Considers "how" the team behaves

Opens communication

Done by consensus

Each team makes its own unique rules

Code of conduct considerations

Commitment of team members. A rule on the amount of participation of team members might be appropriate.

Owners of meeting roles. The rules of conduct may identify the specific meeting roles of the team leader, team members, team facilitator, and meeting recorder.

Negotiation process. A rule for outlining the negotiation process might be appropriate for some teams.

Decision-making process. The process for decision making is a must for most teams.

Unity issues. Rules for maintaining the team's cohesiveness are usually a good idea for the rules of conduct.

Communications procedures. Procedures for allowing all members an opportunity to communicate on all issues are always appropriate.

Time management. Rules for the start and end of the meeting are sometimes needed. Also, rules for conformance to the agenda may be needed by some teams.

Rules of conduct examples

Rely on facts, not opinions.

Understand others' points of view.

Listen actively to all ideas.

Encourage others.

Submit assignments on time.

Open communication of all issues.

Focus.

Come to meetings on time.

Orient toward customer satisfaction.

Never gossip about the meeting or team.

Decide everything by consensus.

Use and build on everyone's ideas.

Conduct the meeting using an agenda.

Take time to self-critique the meeting.

Using the code of conduct

The code of conduct is posted at all team meetings in plain view of all participants, either on a wall or in front of each participant. When a team member notices an issue that is part of the team's code of conduct, she or he simply points to the item on the code of conduct to let the team know. At that time the team takes action to resolve the issue.

As the team moves through the stages of team development, the behavior of the team changes. The team is then required to reevaluate the code of conduct to minimize and manage potential conflicts.

Roles, responsibilities, and relationships

Besides normal team functions, team meetings involve additional roles, responsibilities, and relationships which must be defined. The team leader guides the team to mission accomplishment and may also guide the team during team meetings. Team members are expected to prepare for, participate in, and perform during team meetings. The team facilitator helps the team focus and apply methods, tools, and techniques during the meeting. An additional role for team meetings is that of a recorder. This person prepares all the administration documentation for the meeting, including such items as the agenda, minutes, assumptions, and list of definitions.

In addition to roles and responsibilities, each team member must understand the relationships that exist, as they could affect the team meeting. The relationships involve the team as a whole, other team members, the organization as a whole, the functional organization, and self. A conflict in any of these relationships could cause a team meeting to be canceled or ineffective. These conflicts should be resolved as early as possible to ensure maximum participation by all team members.

Determine team meeting roles

The following are suggested team meeting roles. If your team decides not to use a certain team meeting role, just write "N/A" in the name column. In some teams, each meeting role is assigned to both a primary and an alternate person. In this case, write the name of the primary person first and the name of the alternate second under the name column. It is recommended that each team member have a team meeting role. In addition, detailed minutes need not be taken.

The agenda provides a record of the team's progress. It should be sufficient to inform people outside the team about the team's activities. Typically, team members take sufficient notes to be able to actively participate and to inform other team members.

Role	Primary Name	Alternate Name
Meeting leader		
Recorder		
Assumption recorder		
Glossary recorder		
Chart scribe		
Action item monitor		
Meeting facilitator		

Focus statement

A focus statement provides the purpose of a meeting. Each team meeting must have a written focus statement. If the team cannot write a focus statement, there is no need to hold a meeting. The focus statement should provide the following:

Focus for an entire meeting

Output expected from the meeting

Clear, concise, simple statement

Understanding for everyone on the team

Start for the agenda

Focus statement examples

Informational meeting focus statement: The purpose of this meeting is to *gain insight* into the requirements of an effective meeting.

Action meeting focus statement: The purpose of this meeting is to *create* a mission statement for this team.

Agenda

An agenda acts as the meeting guide. It gets the team to focus on the meeting's desired outcomes. An agenda encourages effective and efficient meetings because it provide a target for the meeting. It documents key team activities and it acts to stimulate progress. An exam-

---

TEAM AGENDA

Date:
Start Time:
End Time:

FOCUS STATEMENT:

ITEMS:

☐   Item:
      Owner:

☐   Item:
      Owner:

☐   Item:
      Owner:

☐   Item:
      Owner:

☐   Item: Next Steps.
      Owner: Team

ACTION ITEM	OWNER	STATUS

Comments:

Figure 6.5   Meeting agenda template.

ple of a meeting agenda is shown in Fig. 6.5. As you can see, it shows exactly what is expected during the meeting. It also provides desired outcomes and action items with follow-up status.

An agenda does the following:

Acts as the team's meeting guide

Gets the meeting's desired outcomes

Encourages effective and efficient meetings

Nurtures a focused team meeting

Documents key team activities

Acts to stimulate progress

Team meetings in action

In addition to the meeting tools mentioned above, the team needs to establish a process for conducting meetings. This process involves actions which must be taken before, during, and after the meeting to ensure proper preparation, conduct, and follow-up. In general, each team member should perform certain functions before, during, and after team meetings, as discussed below.

Before the meeting

The success of the team depends on the active involvement of all team members. Team members should participate fully in all meetings. The following are some meaningful guidelines to assist the team in conducting an effective meeting:

Brainstorm ideas. Review the focus statement and write your ideas of everything you know about the focus.

Evaluate what you know. Start with ideas you brainstormed and gather any additional information you may need. Analyze the information, trying to determine the specific opportunity, problem, or root cause.

Formulate alternatives. Generate a list of alternatives to accomplish the focus.

Orient toward one alternative. Determine one alternative you can support. This is your starting position based on the information you know. During the meeting, you may change your alternative based on additional information provided by other team members.

Review agenda to ensure that you are prepared with information, status, or assignments.

Ensure that you complete any assignments. The team depends on you to accomplish your specific actions. Even if you cannot make the meeting, try to make sure that your assignments are on time.

During the meeting

During the meeting, the key activities of all team members are speaking, listening, and cooperation. Speak to make your point; present and clarify ideas. Listen actively, and cooperate with all other team members. During the meeting, do the following:

Display teamwork

Understand the viewpoint of others

Remain focused

Involve yourself

Nurture others' ideas

Go for win-win solutions

During the meeting...speak

During the meeting, speak to share information but be short, simple, and concise. Plan what you are going to say before you say it. This preparation will help you focus and save the team time. Encourage the building of ideas, which will stimulate interest and involvement. Although you or others may not have anything to initially contribute, many people can add their ideas to others. Avoid personal remarks. Remember, it is not personal. Never use "red" remarks which include any reference to race, creed, culture, family relations, stereotypes, and so on. Keep remarks focused on the mission, goal, problem, and issue. Also, avoid any words that may trigger an emotional reaction such as those that refer to race, sex, religion, politics, and so on.

In summation, do the following when you speak:

Share information, but be concise.

Plan what you are going to say before you say it.

Encourage the building of ideas by asking questions.

Avoid personal or "red" remarks.

Keep remarks focused on subject.

During the meeting...listen

Again, listening is essential during a meeting. Let the other person convey his or her message. Do not interrupt that person while he or she is speaking. Involve yourself in the message. Look for ideas you can support. Determine the central theme or concepts. Summarize and paraphrase frequently. These actions provide the speaker with feedback on the success of the communication. They are also the only way to confirm one's understanding of the information. Further, the summation and paraphrasing may help another team member understand who may not already do so. All critical ideas must be repeated by another member and discussed to ensure clarity of ideas necessary for consensus decision making. Talk only to clarify while you are listening. Effective listening requires your full concentration. Empathize with other people. In other words, put yourself in their shoes for awhile. You do not have to sympathize with them. Empathy helps you understand; sympathy may actually be a barrier. Nurture active listening skills. Active effective listening is not natural. It requires dedicated concentration of effort.

In summation, do the following when listening:

Let the other person convey his or her message.

Involve yourself in the other person's message.

Summarize and paraphrase frequently.

Talk only to clarify.

Empathize with other people.

Nurture active listening skills.

During the meeting...cooperate

Cooperation makes a meeting work. Consider the self-esteem of others, which will give them the confidence to participate. Operate with the team; give others a fair chance. Do not go outside the team to seek action or talk about other team members. Observe others' reactions, which will provide feedback on true reactions. Use the observations to find common ground for negotiations for win-win solutions. Pursue a common focus. As long as the team focuses on a common goal, it can work. Many times a common focus overcomes conflicts as peer pressure to achieve a shared result overshadows the personal needs of team members. Establish open communications, which are necessary for any cooperative effort. Recognize individual contributions, which will help to stimulate more participation. Allow positive conflict, which will lead to consensus decision making. The team will support a decision better if positive debate has been endorsed during the meeting. Trade off ideas with the group, which will distribute ownership to the whole team. Encourage trust. This ingredient is the most important one in developing and keeping cooperation in the team. Without trust, there can be no real cooperation during the team meeting.

In summation, do the following when fostering cooperation:

Consider the self-esteem of others.

Operate within the team.

Observe others' actions.

Pursue a common focus.

Establish open communications.

Recognize individual contributions.

Allow positive conflict.

Trade off ideas of the team.

Encourage trust.

After the meeting

Once the meeting is over, the real team actions are performed. At this time the team members act to perform assignments and action items. Finding support and resources may be necessary after the meeting. A team member coordinates with management or a support function to ensure that the team can complete actions or implement a solution. All team members must talk up team activities to develop pride for their team in the organization. This talk gives all team members a feeling of belonging to a worthwhile team, helps to promote teamwork throughout the whole organization, and provides a means of maintaining team integrity. Team members must not gossip about team activities or other team members. Finally, team members must review the agenda of the next meeting to start preparation for it before it begins.

In summation, team members must do the following after the meeting:

Act to perform assignment(s).

Find necessary support and resources.

Talk up team activities.

Ensure team integrity.

Review the next meeting's agenda.

Decide the meeting process

It is important to decide how the meetings will be conducted. For instance, some teams have the team leader prepare the agenda before the meeting. In other teams, the team prepares the agenda for the next meeting at the end of the current meeting. For each of the "what" items, mark *B* for before the meeting, *D* for during the meeting, and *A* for after the meeting in the "when" column. Assign responsibilities in the "who" column.

What	When	Who
Meeting notice		
Focus statement		
Agenda		
Code of conduct		
Assign action items		
Make decisions		
Monitor progress		
Meeting critique		
Perform action items		
Get resources		
Escalate issues		

Meeting critique

Some teams find it useful to perform a meeting self-assessment at the end of the meeting. This effort is particularly beneficial when a team is just being started. It provides a means to develop the skills required for effective team meetings while also fostering teamwork through the finding of successes in non-mission-related activity work. The more successes a team has as a team, the easier the team can develop and maintain teamwork. The team needs to design its own meeting critique. This critique should be completed as a team at the end of the meeting, which should take no longer than 5 to 10 minutes. The critique should address the following:

Communications - Was there open and honest communication?

Results - Was the focus statement accomplished?

Involvement - Did everyone participate?

Training - Does the team require any training?

Individuals - Were individual contributions recognized?

Questions - Are there any items requiring further research?

Unity - Did the team work together? Are there any symptoms of conflict?

Escalate - Are there any issues requiring management help?

Brainstorming

Brainstorming is a technique used by a group of people that encourages their collective thinking power to create ideas. The purpose of brainstorming is to stimulate the generation of ideas. It adds to the creative power of the team. The value of brainstorming lies in the fact that there may be more than one way to look at a problem or handle it. Through brainstorming, not only are individual ideas or thoughts brought out but they may also spark new ideas or thoughts from others or improve on an idea already under consideration. The more ideas a team has, the greater the probability of finding an opportunity and/or solution.

Brainstorming does the following:

Brings out the most ideas in the shortest time

Reduces the need to give right answers

Allows the group to have fun

Increases involvement and participation

Nurtures positive thinking

Solicits varying ideas and concepts

Tempers negative attitudes

Omits criticism and evaluation of ideas

Results in improved solutions

Maximizes the attainment of goals

Brainstorming rules

For brainstorming to work effectively, the group leader must make sure the principles of brainstorming are followed. Thus, each member must know the rules and follow them.

It is a good idea to review the rules before each meeting until the group has established its brainstorming approach. They are as follows:

Record all ideas.

Use freewheeling ideas.

Limit judgment until later.

Encourage participation by everyone.

Solicit quantity.

Let's look at each of these rules in more detail.

1. *Record all ideas.*   Team members learn over and over the importance of recording things. It is the only way you can recapture what has happened. With brainstorming, it is easy in the excitement to be careless about recording ideas. Be sure someone is appointed to see that everything is recorded.

   Remember, do not allow judgment on ideas during the recording process by letting the recorder omit any ideas. It is best to display every idea in full view of all members on a flipchart, whiteboard, or similar device. After the brainstorming session, all ideas should be recorded on a sheet of paper so the ideas can be preserved for use at a following meeting.

2. *Use freewheeling ideas.*   Freewheeling has value in that while an idea may be unsuitable in itself, it serves as a stimulus for other members of the group. Even wild or exaggerated ideas have thought-provoking value that should never be underestimated.

3. *Limit judgment until later.*   Keep the ideas flowing. No criticism should be allowed, as it will shut off the flow. All ideas are encouraged and accepted. Remember, there are only right ideas. All ideas are right in each individual's mind.

4. *Encourage participation by everyone.*   Good ideas are not necessarily in the minds of a few individuals. Give each member a turn to speak; don't miss anyone. It's important to give ample time for each member to speak.

   For example, solicit responses clockwise around the room. If a member has no idea at the moment, she or he says "pass." By this remark, there is added assurance that no one is missed. Furthermore, a team member that passes on one round may very well have an idea on the next round.

   Encourage participation by building on other ideas.

5. *Solicit a large number of ideas.*   Ideas build on ideas. They can be a combination or extension of other ideas. Ideas are thought-provoking and stimulating. Work toward a large number of ideas. Postpone judgment on ideas; that comes later.

Brainstorming steps

The steps to brainstorming are as follows:

1. *Generate ideas.*   Follow the rules given above.

2. *Evaluate ideas.*   During the evaluation step the team examines

each idea for value. This is the point at which to offer constructive criticism or analyze the ideas presented. Again, it is important that only the idea and not the generator of the idea be criticized. The ideas and alternative combinations of ideas are compared and examined. At this time, some ideas may be eliminated or combined with other ideas.

3. *Decide using consensus.*   Use consensus decision-making techniques as described later in this chapter.

Brainstorming methods

There are three primary brainstorming methods:

1. Round robin
2. Freewheeling
3. Slip

Each method has advantages and disadvantages that the group or discussion leader will have to weigh before determining which one would be best to accomplish desired results. In some cases, the best method may be a combination of the various brainstorming methods. For instance, the brainstorming session may start with a round-robin or slip method and move into a freewheeling method to add more ideas.

Round robin

With this method, each group member, in turn, contributes an idea as it relates to the purpose of the discussion. Every idea is recorded on a flipchart or board. When a group member has nothing to contribute, he simply says "pass." The next time around, this person may offer an idea, if he wishes, or pass again. Ideas are solicited until no one has anything to add.

Round-robin advantages

- It is difficult for one person to dominate the discussion.
- Everyone is given an opportunity to participate fully.

Round-robin disadvantages

- People feel frustration while waiting their turn.
- Ideas are not spontaneous.

Freewheeling

With this method, each team member calls out ideas freely and in a random order. Every idea is recorded on a flipchart or board. The process continues until no one has anything else to add.

Freewheeling advantages

- It is spontaneous and there are no restrictions.
- Many ideas come up in a short period of time.

Freewheeling disadvantages

- Some individuals may dominate.
- A quiet team member may be reluctant to speak.
- It is chaotic if too many people talk at the same time.

Slip method

Each team member writes all her ideas on an issue, a problem, or an alternative on a piece of paper. She writes as many ideas as possible. Then the slips are collected and all the ideas are written on the board. A variation to this method is the Crawford slip method where each idea is written on a separate slip of paper. The slips are then put on a board and arranged in categories.

Slip-method advantage

- All ideas are recorded and all contributions are anonymous.

Slip-method disadvantage

- Some creativity may be lost due to the inability of the other team members to react to the contributions of others.

Brainstorming example

Figure 6.6 provides an example of a brainstorming session on the barriers to teamwork.

Advanced brainstorming techniques

There are many advanced brainstorming techniques beyond the basic three mentioned above. Two of the most popular of these advanced techniques are nominal group techniques and affinity diagrams.

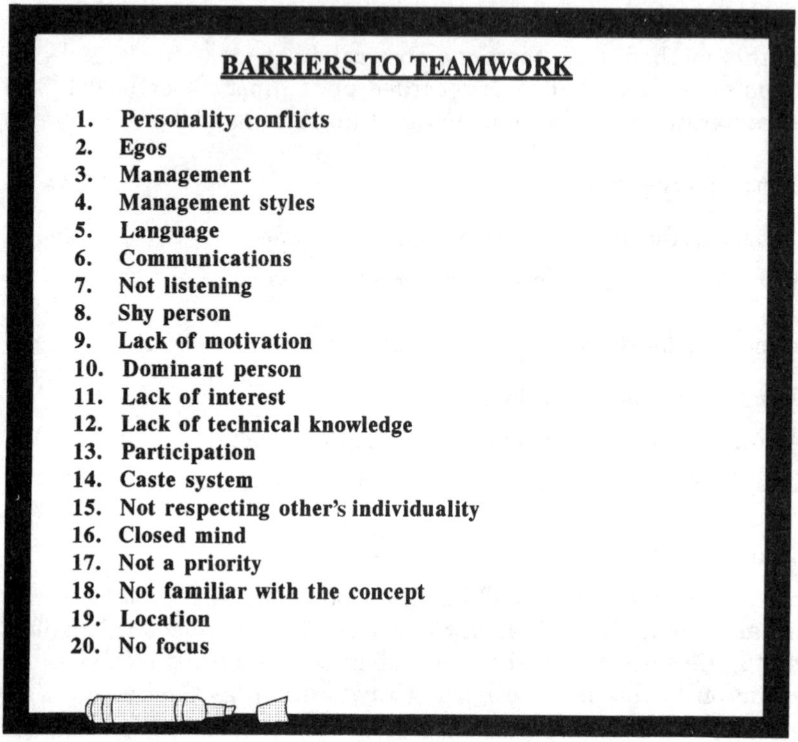

BARRIERS TO TEAMWORK

1. Personality conflicts
2. Egos
3. Management
4. Management styles
5. Language
6. Communications
7. Not listening
8. Shy person
9. Lack of motivation
10. Dominant person
11. Lack of interest
12. Lack of technical knowledge
13. Participation
14. Caste system
15. Not respecting other's individuality
16. Closed mind
17. Not a priority
18. Not familiar with the concept
19. Location
20. No focus

Figure 6.6   Brainstorming example.

Nominal group technique

Nominal group technique is a refinement of brainstorming. It provides a more structured discussion and decision-making technique. The nominal group technique allows time for individual idea generation. This amount of time granted may vary. Sometimes, if the subject is not too complex, the team may only have 5 to 10 minutes. For a complex issue, team members may be asked to generate their ideas between team meetings. Once the ideas are generated, the nominal group technique then allows the leader to survey the opinions of the group about the ideas generated. Finally, nominal group technique leads the group to set priorities and focus on consensus. The nominal group technique steps can be summarized as follows:

1. Present the issue and give instructions.

2. Allow time for idea generation.

3. Gather ideas via round robin, one idea at a time. Write each idea on a flipchart or board and post.

4. Process or clarify ideas. Focus on clarification of meaning, not on arguing points. Eliminate duplicate ideas; combine similar ideas.

5. Set priorities.

Affinity diagram

The affinity diagram is another idea generator. It starts with the issue statement. Once the issue is presented, it continues like the nominal group technique with some time allowed for individual idea generation. The difference with an affinity diagram is that each of the ideas are written on an index card or adhesive note commonly known as a Post-it. Each idea is recorded by the individual on one index card or adhesive note. All notes are then posted on a wall or put on a table. The team members then put the index cards or adhesive notes into similar groupings, doing so without discussion. Next, the team, through discussion, decides on a theme for each group of notes. The team creates a header card for each group of notes from the theme. The cards are arranged under the issue with the header card on top and each of the notes relating to the grouping underneath the header card. Next, the items are prioritized for action just as with the nominal group technique.

Consensus decision making

Consensus decision making means that everyone on the team accepts and supports the decision. This does not mean that everyone wants the same decision but that everyone on the team agrees to go along with the decision. Consensus equals commitment to the decision. It can only be reached by open and fair communication among all team members. Consensus is critical when the team is developing a code of conduct, vision, mission, charter, and values, or when it is selecting a process to improve, a problem to solve, a mission to accomplish, or an opportunity to pursue, along with accompanying recommendations for courses of action and/or solutions. Consensus requires understanding and discussion among all stakeholders. Once understanding and discussion take place, the group or team can proceed with the process of arriving at a consensus decision.

Communicate.

Open team members' minds to new ideas.

Nurture ideas of all team members.

Share information.

Encourage everyone's participation.

Nurture active discussion; don't vote.

Support ideas that are best for everyone.

Understand that differences are strength.

Seek win-win solutions.

When to use consensus decision making

Consensus decision making is the most desirable decision-making method when there are critical decisions requiring commitment and support that need to be made. However, it is not always the most desirable decision-making method. In making a decision, the timing, significance, and necessary support of the outcome should be considered. A consensus decision takes time. However, a group will be more committed to success if the decision is reached by consensus. Consensus decision making targets a win-win outcome. Decisions reached by any method other than consensus can result in a win-lose situation. A win-lose decision equates to not having total commitment and support for the selection. Therefore, it is important to use consensus decision making when seeking total commitment for organization-wide results. For instance, consensus decision making should be used when one is

Developing a focus (vision, mission, charter, goal)

Formulating a code of conduct

Selecting an issue, problem, or opportunity

Deciding on a solution to implement

Other methods of decision making also exist. These types of decision-making methods may be necessary at times when consensus decision making is not appropriate. In many instances, time constraints, insignificance of the decision, or other considerations make consensus decision making unrealistic. In certain situations, the organization uses an alternate method of decision making from the methods below:

Decision by majority. This is a decision by more than half of the representatives.

Decision by leader or minority. In some cases, the leader or owner makes the decision.

Decision by management. Management sometimes must make the decision.

Consensus decision-making action process

Consensus is reached by allowing everyone the opportunity to express their ideas about a decision. The action process for reaching a consensus decision are as follows:

1. *Present the decision to be made.* The decision statement should include the what, when, and why. The decision statement should not be given as an either/or alternative.

2. *Write the decision statement along the top of a flipchart.*

3. *Review background information.* Provide a common foundation of information to all participants or team members.

4. *Decide how the decision should be made.* Conduct a discussion to determine if consensus is the best decision method for this situation. Review the considerations for consensus decision making.

5. *Brainstorm selection criteria.* First, each team member takes five minutes to write items for selection criteria. Second, the team conducts a round-robin brainstorming session to list items on the flipchart. This stage involves each team member, in turn, providing one item from her or his sheet of paper. This task is continued until all items from the individual sheets of paper are listed on the flipchart. Third, items are added to the list during a free-wheeling brainstorming session. During this stage the list is opened to everyone to contribute.

6. *Clarify ideas.* Discuss each item requiring explanation.

7. *Agree on selection criteria for the decision.* The team needs to decide which items should be included in the decision selection criteria. Command media (laws, regulations, policy, etc.) should be considered as essential selection criteria.

   - Evaluate each item in terms of a yes, no, or maybe for criteria.
   - Resolve the maybe items.
   - Add command media items to list.

8. *Brainstorm alternatives.* Write the alternatives on a flipchart.

9. *Evaluate each alternative against selection criteria.* In some cases, it may be necessary to use selection techniques to focus alternatives to a manageable number (between 2 and 5). Selection techniques include voting, selection matrix, and selection grid.

10. *Agree on a decision.* Write the final decision on a flipchart and post.

11. *Get each team member's personal commitment to the decision.* Ask each team member individually whether she or he agrees with the decision.

12. *Implement the decision.* Develop an action plan with the what, when, and who to implement the decision. Conduct periodic reviews to follow up.

Presentation

Sometimes a presentation may be necessary to provide information, obtain approval, or request action. The presentation may be formally or informally given by the team. Involve as many team members as possible in the actual presentation. The presentation provides the opportunity to inform everybody of team activities and accomplishments and recognize team members for their contributions.

Presentation steps

Step 1: Gain support. Gaining support requires identifying and involving key people early in the improvement process. Ensure support for recommendation from owners, suppliers, and customers by stressing benefits to the organization.

Step 2: Prepare the presentation

- Anticipate objections.
- Rehearse the presentation.
- Arrange the presentation.

Step 3: Give the presentation

- Build rapport.
- Make the recommendation.
- Stress the benefits.
- Overcome objections.
- Seek action.

Step 4: Follow up on the presentation

- Follow up to ensure that the recommended action is implemented.
- Reduce postdecision anxiety by repeating and summarizing benefits.
- Stress the benefits of early implementation.

Prepare the presentation

Once the team knows they have sufficient support for a recommended course of action that requires management approval, they must prepare the presentation. Preparing the presentation involves the activities as listed above. To accomplish these activities, the following processes must be performed:

- Develop presentation materials.
- Produce the presentation materials.
- Arrange for the presentation.
- Practice the presentation.

Develop presentation materials

Development of presentation materials involves developing a specific objective for the presentation and preparing a presentation outline to accomplish the objective. The presentation objective should state specifically the expected outcome of the presentation. The objective should be stated in terms of who, what, and when.

Presentation objective example

The organization development and training manager will analyze within three months the specific needs of the organization to implement customer-driven quality improvement teams throughout the organization.

In this example, the who is the organizational development and training manager. The what is analyzing the specific needs of the organization to implement customer-driven quality improvement teams throughout the organization. The when is within three months.

Presentation outline

The outline of the presentation should be geared to accomplish the objective. When preparing the presentation outline, consider the audience, understand how the recommendation affects others, and outline the organizationwide benefits. The audience may be supportive or unreceptive. Conduct a force-field analysis to determine the restraining forces and driving forces of the audience. At the same time, consider how the recommendation will affect others. Anticipate objections. Again, conduct a force-field analysis to determine driving forces of any known objections to your proposal. Further, outline the organizationwide benefits through brainstorming and data collection.

Now you are ready to prepare the presentation outline. The presentation outline should contain an introduction, body, and conclusion. In the introduction, tell them what you are going to tell them. In the body, tell them. Then, in the conclusion tell them what you have just told them.

I. The introduction
   A. Establish rapport with introductions.
   B. Get the audience's attention by giving benefits.
   C. Tell them what you are going to tell them.
II. The body
   A. State your mission.
   B. Describe the process using a process diagram.
      1. Significance of the process
      2. Inputs with suppliers
      3. Process itself
      4. Output(s) with customer(s)
      5. Owner(s)
      6. Identify the underlying cause
      7. Describe data collection
      8. Discuss results
   C. Detail the action requested.
      1. Alternatives considered
      2. Solution selected
      3. Plan for implementation
III. The conclusion
   A. Reinforce benefits.
   B. Tell them what you told them.
   C. Get agreement on what you want.
   D. Summarize actions.

Prepare presentation materials

Presentation materials can be as simple or as complex as required to get the requested action from the audience. They attract and maintain attention on main ideas; illustrate and support the team's recommendations; and focus on minimizing misunderstanding. Presentation materials could include handouts, overhead transparencies, flipcharts, videotapes, and computer-based visuals. As a minimum, the presentation material should consist of a handout for all participants. Normally, the presentation materials consist of a handout and some form of visual aid for the group to observe, usually either a flipchart or overhead transparencies. Specific tips for preparing the most common presentation aids of handouts, flipcharts, and overhead transparencies are given below.

Handouts. A handout supports the presentation by providing critical information and/or supplemental detail. If the handout is distributed prior to the presentation, it should follow the outline of the presentation. If the handout only provides supplemental or reinforcing information, it should not be provided to the audience until an appropriate time during the presentation or at the conclusion of the presentation.

Flipcharts. Flipcharts enhance the presentation. They should emphasize the key points or graphically show concepts. Some specific tips for the design of flipcharts are

- List main points as bullets.
- Limit bullets to six or less per chart.
- Keep bullets short—around six words or less per bullet.
- Make sure that chart is readable from every seat in the room.
- Use multiple colors to stress key words.
- Leave a blank sheet between flipcharts.
- Have charts reflect the professional pride of the team.

Overhead transparencies. Overhead transparencies are the most frequent material used to aid in the understanding of information in a presentation. They function in the same way as flipcharts. However, with today's computer technology, especially graphic capability, overhead transparencies should be used to reinforce ideas graphically as well as verbally. Studies have shown that people have different styles for understanding, depending on which side of the brain dominates—left or right. So it may be appropriate to use both words and pictures to convey your message. The words appeal to the more logical, left-brain-dominant people, and the graphics appeal to the creative right-brain-dominant people.

The tips for flipcharts apply as well to overhead transparencies. But take care not to attempt to show large amounts of data or complex processes on one overhead. Reduce all the information to show just trends, relationship, or overall processes. If detailed information is necessary, provide this material in a handout.

Produce the presentation material. Depending on the situation, the team may produce the presentation materials itself or it may have to rely on various support services to produce them. If the team produces the materials, ensure that the presentation meets the standards of the audience. If the team has support services to make the presentation materials, planning and coordination are important. Many organizations have support services such as word processing, editing, graphics, and printing. When using these services, the team must plan enough time to allow for professional workmanship. They must determine all the tasks to be accomplished and allow an appropriate time period to meet the team's scheduled presentation date. Enough time should be allotted to use the presentation materials for a dry run before the actual presentation. Also it is wise to periodically coordinate with the support services people to ensure progress toward meeting the schedule.

Arrange for the presentation

Administration details can have an effect on how the presentation is received. Ensure that the following administration details are accomplished:

- Schedule presentation time and place.
- Ensure that all the right participants can attend.
- Set up the room.
- Have presentation materials.

Practice the presentation

Rehearse the presentation prior to the actual presentation. If possible, practice the presentation one time to an audience that provides a representation of the actual audience.

Give the presentation

Giving the presentation involves presenter preparation and the actual conducting of the presentation. The best way to ensure a successful presentation is by adequate preparation, which is enhanced through the development of some basic presentation skills. These skills can be grouped in the following categories:

- Presenter's preparation
- Presenter's style
- Presenter's delivery

Presenter's preparation

Depending on your experience, you will be more or less comfortable presenting to a group. Your level of comfort can be improved with time spent preparing. Specifically, the following techniques can be used to enhance your comfort level:

- *Practice.* Practice the presentation enough to get very familiar with the flow of ideas. Do not memorize the presentation. Practice enough to allow you to present the information naturally.

- *Plan for objections.* Perform a force-field analysis to determine objections and your response.

- *Visualize success.* Prior to the presentation spend some time alone picturing yourself accomplishing a successful presentation.

Presenter's style

Although each presenter has his or her own style, the following guidelines will help any presenter become more successful:

- *Act naturally.* Make the presentation as natural as possible. Try to avoid doing anything that would appear faked, forced, or flaky.

- *Maintain positive attitude.* Display a positive attitude by showing enthusiasm. Above all, be sincere in your commitment and support for the presentation goal.

Presenter's delivery

There are several nonverbal and verbal presentation tips to improve anyone's presentation. Most people concentrate on the verbal communication aspects of the presentation, although nonverbal behaviors such as eye contact, body movement, and gestures communicate much of the meaning. Nonverbal communication skills enhance your ability to effectively communicate to the audience.

- Eye contact shows interest in the audience. Look directly at your audience and include everyone equally. Good eye contact results in enhanced credibility.

- Body movement is another important physical behavior for a presenter. It helps hold the audience's attention and puts the speaker at ease by allowing her or him to work off excess energy that can cause nervousness. You can use body movement as punctuation to mark a change in your presentation. Moving from one spot to another tells the audience you are changing the line of thought. Some body movement can be distracting. Pacing back and forth, rocking from side to side, or "dancing" serves no purpose and tells the audience that you are nervous.

- Gestures can clarify, emphasize, or reinforce what is said. Make gestures by using your hands, arms, shoulders, and head. Fidgeting with your watch and scratching your ear are not gestures. This type of behavior usually distracts from the presentation. Practice is needed to use gestures effectively.

Conduct the presentation

During the presentation, the team does the following:

- Builds rapport by developing a friendly but professional relationship with the audience.
- Makes the recommendation using the results of the Total Quality Management improvement methodology. Support the recommendation with facts.
- Stresses the benefits of implementing the team's recommendation. The benefits should emphasize the tangible measurable gains of the solution. In addition, intangible advantages should be shown.
- Overcomes objections by using the driving forces driven from the force-field analysis. Remember, focus on the issue; never make it personal.
- Seeks action for implementation. The conclusion must provide a definite course of action.

Follow up on the presentation

After the presentation, the team should do the following:

- Follow up to ensure that the recommended action is implemented.
- Reduce postdecision anxiety by repeating and summarizing benefits.
- Stress the benefits of early implementation.

Improving Continuously

Improving continuously involves using the continuous improvement cycle and improvement methodology with appropriate tools and techniques.

Continuous improvement starts at the organizational level through use of the continuous improvement cycle. Within the continuous improvement cycle, an improvement methodology must be used for process improvement and problem solving. Such use will foster preventive action and problem solving. Figure 7.1 shows the continuous improvement cycle.

The continuous improvement cycle is the overall cycle for improvement. This cycle targets and is used mainly for top-level improvement activities. The improvement methodology outlined in this book is useful for any organization. It employs commonly understood problem-solving techniques for both problem solving and process improvement. At the working level, the improvement methodology is the only methodology used for continuous process improvement. The basic problem-solving methodology instructs the user to understand, select, analyze, generate alternatives, select a solution, plan and gain approval, institute, and check.

There are many different improvement methodologies being used by various organizations. Some organizations use statistical process control; others use quality function deployment; and still others focus entirely on process analysis. Again, the improvement methodology must be geared to the specific organization. No matter what improvement methodology the organization decides to use, those procedures should be used throughout the entire organization. Such organizationwide employment will assist in the creation and maintenance of a systematic, integrated, and consistent perspective. The improvement methodology recommended by this book has been selected because it

Figure 7.1 Continuous improvement cycle with improvement methodology.

is commonly understood by a wide variety of people and it is simple to use for any organization.

Basic Improvement Methodology

A basic improvement methodology is outlined below. Steps 1 through 8 comprise a basic problem-solving methodology which can be repeated continuously.

1. Understand/identify the opportunity.
2. Select an opportunity for improvement.
3. Analyze the selected opportunity.
4. Generate improvement alternatives.
5. Select an improvement alternative.
6. Plan and gain approval for the selected improvement.
7. Institute the selected improvement.
8. Check results.

Using the basic improvement methodology for process improvement

The basic improvement methodology is used throughout the entire organization for action planning, problem solving, and process improvement. It can be modified as appropriate for each specific application.

Figure 7.2 shows the basic improvement methodology modified for process improvement. The steps include understanding the process, selecting a critical process, analyzing the critical process, generating improvement alternatives, selecting an improvement, planning and gaining approval, instituting the improvement, and checking the results.

In cases where an improvement would have a significant impact, inserting steps 5a to 5d is recommended. These steps are used to test or pilot an alternative to determine if it will provide desired outcomes. Use of these procedures prevents one's having to go through all the steps to institutionalize the improvement. Steps are as follows:

5a. Plan the improvement on a test or pilot basis.

5b. Do the improvement on a test or pilot basis.

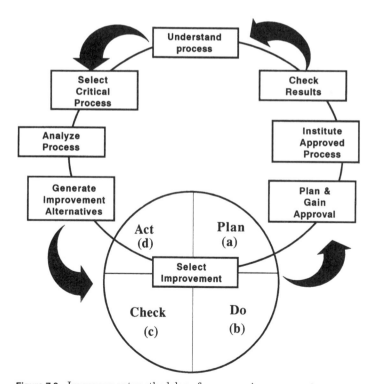

Figure 7.2   Improvement methodology for process improvement.

5c. Check results of test or pilot to desired outcome.

5d. Act to make the improvement permanent by going to step 6, or repeat starting with 5a or going back to step 1.

Be aware that in most cases insertion of these steps is not appropriate. If it is not, continue with step 6.

Using the improvement methodology for process improvement steps

The following is a step-by-step detailed outline for using the improvement methodology for process improvement.

Step 1: Understand the process.   During this first step, focus all efforts on understanding the process's specific contributions to customer satisfaction. Specifically, to fully understand the process do the following:

- Define the overall process.
- Diagram the top-level and top-down process.
- List the customers' needs and expectations from the process.
- Determine if the process is meeting the customers' expectations.
- Discover who owns and influences the process.
- Determine all the inputs and outputs to the process.
- Understand the relationship between inputs and outputs.
- List the suppliers of the inputs.
- Determine if the suppliers are meeting the requirements.
- Specify the customer(s) of the process.
- Determine how to measure or use metrics for the process.
- Measure the process to determine how it is performing.
- Understand the value of the process to the product or service.
- Determine if the process can be eliminated.
- Benchmark process.
- List the problems with the process as it exists.

Step 2: Select a critical process for improvement.   Select a critical process for improvement based on the information collected in the understanding stage. A critical process usually is a process that is not meeting the expectations of the customer. To select a part of the process for improvement, do the following:

- List areas of the process requiring improvement, including those that are not meeting the needs of the customer, those in which inputs are not meeting requirements, and problem areas.

- Specify the selection criteria.

- Determine the selection method.

- Define how the decision will be made.

- Make the decision.

Step 3: Analyze the selected critical process. This step requires a thorough use of analytical tools to focus on process variation or underlying causes of process problems. When analyzing the selected process part, do the following:

- Determine if the whole process can be eliminated.

- List the detailed steps in the process.

- Diagram the process.

- Look for ways to eliminate non-value-added steps or simplify the process.

- Eliminate or reduce wait times.

- Remove any unnecessary loops.

- Decrease any complexity.

- Analyze frequency changes.

- Eliminate or reduce any waste.

- Look for other ways the process or work can be done.

- Find any problem areas.

- Determine underlying cause.

- Ask five whys.

- Define measurements.

- Find or collect data.

- Complete data gathering.

- Organize data.

- Define the expected outcomes (goals).

- Determine if the process is meeting the goal(s).

- Analyze the forces at work in the situation.

- Determine restraining forces.

- Specify driving forces.

Step 4: Generate improvement alternatives. During this step, the use of creativity, innovation, and imagination is encouraged to explore as many improvement alternatives as possible, including improving the current process, reengineering the process, or inventing a new process. When generating improvement alternatives, do the following:

- Define alternatives that can be used to reach the goal.
- Determine all the forces at play with each alternative.

Step 5: Select an improvement. Select an improvement most likely to attain your desired outcome. If possible, complete the *p*lan, *d*o, *c*heck, and *a*ct (PDCA) cycle and run a test or pilot of selected improvement. Plan the pilot or test; do the improvement on a test or pilot basis; check results against desired outcome; and act to make the improvement permanent. Specifically, do the following:

- Specify selection criteria.
- Define the selection method.
- Determine how the decision will be made.
- Make the decision.
- Complete the PDCA cycle for the selected alternative, if necessary and feasible.

Step 6: Plan and gain approval for improvement. This step requires the preparation of a complete implementation plan. Further, a presentation may be needed so that approval for implementation of the improvement may be obtained. During this step, do the following:

- Determine how the improvement will be implemented.
- Prepare an implementation plan.
- Gain support.
- Stress benefits.
- Determine if a presentation is required.
- Prepare the presentation.
- Request action in the presentation.
- Overcome objections during the presentation.
- Follow up on actions after the presentation.

Step 7: Institute the improvement. This step involves institutionalizing the improvement by installing a feedback system, developing procedures, and/or providing training. Specifically, do the following:

- Install a continuous feedback system.
- Develop, document, and implement procedures.
- Provide training.

Step 8: Check the results for the desired outcome. At this step, a continuous check of results is needed to ensure that the process stays under control. If the process is not meeting desired outcomes, return to steps 1, 3, and 4 (understand, analyze, and select alternatives steps). Repeat all of the steps, 1 through 8, as many times as necessary to achieve the overall goals. During this step 8, do the following:

- Measure the performance against the expected outcomes (goals).
- Determine if you are meeting those goals.
- Continue to keep the process under control.
- Continuously improve the process.

Using the basic improvement methodology for problem solving

The basic improvement methodology for problem solving, as shown in Fig. 7.3, is as follows:

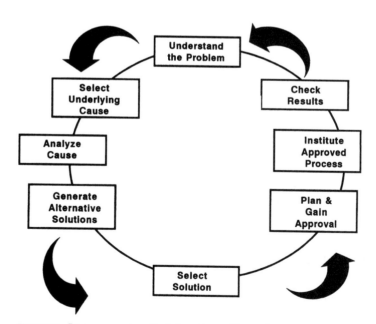

Figure 7.3 Improvement methodology for problem solving.

1. Identify the problem.
2. Select the underlying cause of the problem.
3. Analyze the problem.
4. Generate alternative solutions.
5. Select a solution.
6. Plan and gain approval.
7. Institute the solution.
8. Evaluate the outcome.

Improvement Tools and Techniques

An improvement methodology in which appropriate tools and techniques are used is an essential element of Total Quality Management. When an orderly approach is used, as it is in the continuous improvement cycle with improvement methodology, processes are improved and problems are solved. It is important for everyone to know where they are going and how they are going to get there. The continuous improvement cycle provides this methodology.

Equally important is using proven tools and techniques within the continuous improvement cycle and improvement methodology. Later in this chapter, some areas where specific tools and techniques may be effective within improvement methodology are shown. However, each organization and individual is encouraged to use the tools and techniques in any way appropriate to their specific applications. Further, the tools and techniques can be tailored to your specific application. If it works, use it.

There are many tools and techniques for improvement, some of which have been discussed elsewhere (see, for example, Chapter 6). This chapter focuses on the technical tools and techniques for process improvement and problem solving which, for organizational purposes, are arranged in the following categories:

Process understanding

Selection and decision making

Analysis

Process understanding tools and techniques

Process understanding tools and techniques are essential to the improvement methodology; they are benchmarking, process diagrams, input/output analysis, and supplier/customer analysis. The following is a brief description of each tool.

- *Benchmarking* is a method of measuring your organization against those of recognized leaders.

- *Metrics* are meaningful measures that target continuous improvement focusing on customer satisfaction.

- *Process diagrams* are tools for defining the process.

- *Input/output analysis* is a technique for identifying interdependency problems.

- *Supplier/customer analysis* is a technique to obtain and exchange information for conveying your needs and requirements to suppliers and mutually determining needs and expectations of your customers.

Selection tools and techniques

Selection tools and techniques are used several times during the improvement methodology to help clarify assumptions and focus on consensus when an improvement opportunity or an actual improvement is being selected. The selection tools and techniques are voting, selection matrix, and selection grid. The following is a brief description of each tool:

- *Voting* is a technique to determine majority opinion.

- A *selection matrix* is a tool for rating problems, opportunities, or alternatives based on specific criteria.

- A *selection grid* is a tool for comparing each problem, opportunity, or alternative against all others.

Analysis tools and techniques

During the improvement methodology, thorough analysis is extremely important. The tools and techniques for analysis help improve the process, determine underlying causes, identify the vital few, and describe both sides of an issue. The analysis tools and techniques are process analysis, cause-and-effect analysis, the five whys, data collection and analysis, and force-field analysis. These tools are outlined in Chapter 8. The following is a brief description of each tool.

- *Process analysis* is a tool to improve the process and reduce process cycle time by eliminating non-value-added activities and/or simplifying the process.

- *Cause-and-effect analysis* is a technique for helping a group examine underlying causes.

- *Five whys* is a tool to get to the "root" cause quickly.

- *Data statistical analysis* is actually several tools for collecting, sorting, charting, and analyzing data.

- *Force-field analysis* is a technique that describes the forces at work in a given situation.

Use of tools and techniques within the improvement methodology

The TQM tools and techniques can be used in many places within the improvement methodology. The following are recommendations for use of tools and techniques during specific steps in the process improvement methodology.

Understanding the process tools and techniques. The following tools and techniques are useful for understanding the process:

Focus (vision and mission) setting

Process diagrams

Input/output analysis

Supplier/customer analysis

Benchmarking

Metrics

Brainstorming

Selecting critical process for improvement tools and techniques. The following selection tools and techniques help clarify assumptions and focus on consensus when a part of the process for improvement is being selected:

Voting

Selection matrix

Selection grid

Consensus decision making

Analyzing the selected critical process tools and techniques. The following tools and techniques are useful when the selected process part is being analyzed:

Process diagrams

Process analysis

Cause-and-effect analysis

Five whys

Force-field analysis

Focus (goal) setting

Data statistical analysis

Generating improvement alternatives tools and techniques. Use one of the following tools and techniques when generating alternatives:

Brainstorming

Force-field analysis

Selecting improvement tools and techniques. The following selection of tools and techniques help clarify assumptions and focus on consensus when an alternative is being selected:

Voting

Selection matrix

Selection grid

Consensus decision making

Planning and gaining approval improvement tools and techniques. The following tools and techniques assist with planning and gaining approval for the selected improvement:

Force-field analysis

Presentation

Instituting the selected improvement tools and techniques. Install a continuous feedback system, develop document and implement procedures, and provide training to implement the selected improvement. Project management tools and techniques are useful for ensuring implementation.

Checking the results for the desired outcome tools and techniques. The following tools and techniques help check the results for the desired outcome:

Data/statistical analysis

Supplier/customer analysis

Metrics

8

Improving Processes:
Tools and Techniques

Improving processes involves understanding, selection, analysis, generating improvements, gaining approval and support, instituting, and checking results.

This chapter focuses on the following tools and techniques:

1. Process understanding tools and techniques:
   - Process diagrams
   - Input/output analysis
   - Supplier/customer analysis
   - Benchmarking
   - Metrics

2. Selection tools and techniques:
   - Voting
   - Selection matrix
   - Selection grid
   - Decision making

3. Analysis tools and techniques:
   - Process analysis
   - Work-flow analysis
   - Five whys
   - Cause-and-effect analysis
   - Data statistical analysis
   - Force-field analysis

4. Implementation tools and techniques:
- Task list
- Project schedule

Process Improvement

A fundamental principle of TQM is continuous improvement of processes. Continuous process improvement is the method to meet constantly higher demands from customers while optimizing costs. This effort requires organizations not to simply downsize and reduce but to make radical improvements in efficiency and effectiveness of processes. This chapter provides the tools and techniques to accomplish basic process improvement activities which are the way to improve the work that is accomplished daily.

Process improvement activities are normally accomplished by process improvement teams, quality action teams, and so on. These teams are typically cross-functional and have the mission to improve a process. The tools and techniques discussed in this chapter will help these teams achieve "real" success if used with TQM philosophy and guiding principles.

Figure 8.1 shows the objective of continuous process improvement to ultimately reach excellence through a series of improvements over time.

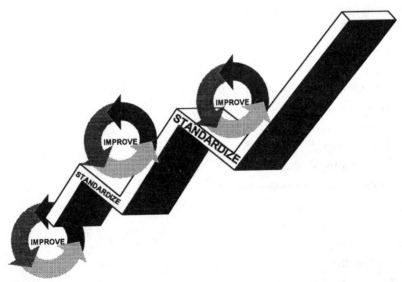

Figure 8.1 Continuous process improvement.

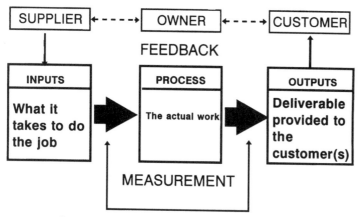

Figure 8.2 Graphic representation of process.

The process

Before process understanding tools and techniques are detailed, the basic process definitions, the nature of a process, the states of a process, process measurement, and process variation must be defined.

Process definitions

The following are some basic definitions required to understand a process:

- A process is a series of activities that takes an input, modifies the input (work takes place and/or value is added), and produces an output. Thus, a process is the job itself. Figure 8.2 presents a graphic representation of a process.
- An input is what you need to do the job.
- An output is the product or service given to someone else.
- A supplier is the provider of the people, material, equipment, method, and/or environment for the input to the process.
- A customer (internal or external) is anyone affected by the product or service.
- The process owner is the person who can change the process.
- Continuous process improvement is the never-ending pursuit of excellence in a process performance.
- A measurement of a process is the difference between the inputs and the outputs of the process as determined by the customer.

■ Variation of a process is any deviation from its ultimate best target value.

Process considerations

People add value.

Requirements/expectations must be communicated.

Ownership develops pride in workmanship.

Continuous improvement targets optimum customer satisfaction and internal processes.

Empowerment leads to unlimited potential.

Supplier partnerships and customer relations are key.

Satisfaction of the customer is always the primary focus.

The hierarchical nature of a process

An understanding of the hierarchical nature of a process as displayed in the top of Fig. 8.3 is important when a process is being improved. There are many levels of processes. At the top level is the major processes. These top-level processes can be broken down into subprocesses. Each subprocess consists of many tasks.

The bottom of Fig. 8.3 shows an example of the nature of a process. At the top level in the figure there are the three major processes in producing a part. A part is engineered, it is manufactured, and it is tested. The manufacturing process breaks down into its subprocesses, another level of processes. When a part is manufactured, a shop order is prepared; material kits are provided; the part is fabricated; and the part is inspected. The kitting subprocess consists of several tasks. The subprocess of building a material kit requires pulling parts, preparing the kit, and releasing the kit to the shop.

Because of the various levels of processes, determining the boundaries of a specific part or parts of a process is a necessity. So the start and finish must be defined. Further, processes impact other processes on the same level or on different levels above or below. Therefore, the impact of any improvement effort on other processes must be known before the improvement is implemented. For instance, if an improvement recommends making a square hole, a further process may have used a round peg. Therefore, the other process would need to be changed to a square peg.

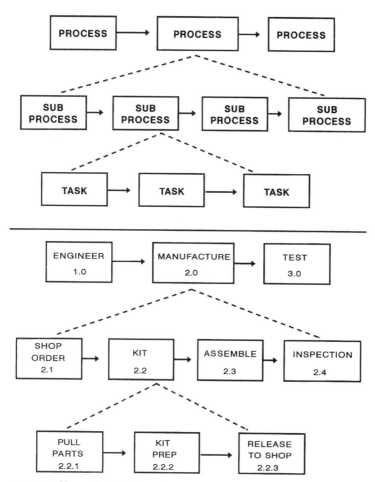

Figure 8.3 Hierarchical nature of process.

Process states

A process can be in one of several states, depending upon variation and capability. Figure 8.4 shows the various states of a process.

State 1 is the unknown state. In this state, the process performance has not been measured. There is no target. State 2 shows the process out of control. There is a target, but the performance cannot be predicted. In this state, the process performance is an element of chance. State 3 displays a process in control, but the process is not capable. The process performance can be predicted, but it will not always hit the target. In this state, the process is not within limits. State 4 is a

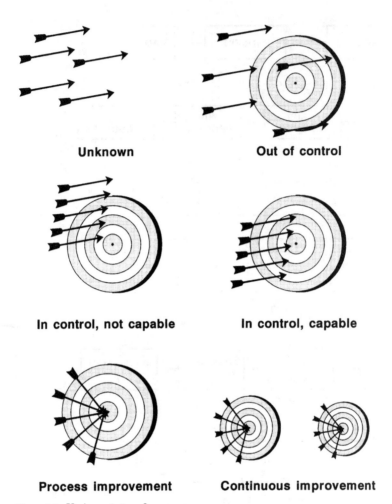

Figure 8.4 Various states of a process.

process in control and capable. The process performance can be predicted within the target. State 5 is process improvement. In this state, the process is improved to reduce variability to the target value. The aim is to consistently hit the bull's-eye or center of the target. State 6 is continuous improvement. In this state, the process is constantly improved to its best possible performance. The target keeps getting smaller and smaller while the bull's-eye is still capable of being hit.

The continuous improvement system moves the process from one state to another with the ultimate aim to consistently perform each process at its ultimate best.

Process measurement

All process performance can be measured through process indicators. The major process indicators focus on quality, cost, quantity, time, accuracy, reliability, flexibility, effectiveness, efficiency, and customer satisfaction. Usually, these process performance indicators are either the difference between the input and the output of the process, or the output of the process from a customer's viewpoint. In manufacturing processes, the process performance indicators are called quality characteristics. These quality characteristics are items such as weight, height, thickness, strength, color, temperature, and density. Besides these quality characteristics, there are process performance indicators in every aspect of the organization. The possibilities of performance indicators is only limited by the view of the organization. Each organization and process owner must determine its own process performance indicators. Some other examples of process performance indicators are errors, time to deliver to the customer, orders filled, number of repairs, number of skilled personnel, response times, number of changes, customer complaints, spares used, mean time between failures, and time available for operational use.

As a minimum, the organization needs to measure the areas critical for business success, key internal business processes, and customer satisfaction. The areas critical for business success are the 3 to 10 most significant indicators of the organization's performance. For instance, each one of the elements in the VICTORY-C model is continuously measured in a never-ending quest. IBM Rochester formulated its plans based on six critical success factors when it won the Malcolm Baldrige National Quality Award. The six areas are improved products and services definition, enhanced product strategy, six-sigma defect elimination strategy, further cycle-time reductions, improved education, and increased employee involvement and ownership. Cadillac uses five "targets of excellence." Federal Express focuses on 12 service quality indicators.

Besides the key areas for business success, each process within the organization needs a process performance indicator. These indicators need to be developed by the process owners and should aim toward continuous improvement. They should not be used to drive or control any of the people in the organization; instead, they should be focused on the process. Some typical process performance indicators of internal processes include cycle time, cost, schedule, number of items, amount of rework, number of errors, number of failures, delivery time, and so on.

In addition to the above areas for measurement, customer satisfaction is an essential measurement area. The development of measurements in this area requires communication with customers to determine the exact measure for total customer satisfaction. Usually customers want deliverables to satisfy their specific needs, at the time they are needed, whenever they are needed, for as long as they are needed, and at a cost they can afford. Throughout this book are many examples of customer satisfaction indicators. These indicators aim at performance, response time, cost, availability, reliability, value, service, use, appearance, and so on.

The results areas mentioned in the Malcolm Baldrige National Quality Award provide an excellent starting point for process measurement. Each organization determines specific process measurements based on the following:

- Product and service quality results
- Company operational results
- Business process and support service results
- Supplier quality results
- Customer satisfaction results

Process variation

There is variation in every process as a result of common and special or assignable causes. Common causes are the normal variation in the established process. They are always part of the process. Special or assignable causes are abnormal variations in the process. They arise from some particular circumstance. It is important to understand the impact of both causes of variation. A variation from a special assignable cause should be solved as a specific problem attributed to something outside the normal process. A variation from a common cause can only be improved by a fundamental change in the process itself. If a common cause is mistaken for a special cause, the adjustment of a common cause could result in increased variation and it will frequently make the process worse. For example, a test failure is encountered when a final test of an assembly is performed. The test failure is determined to be attributed to the "A Board." If the failure's root cause is further blamed on the special cause of operator error in the fabrication of the board when in fact the procedures in the process were incorrect, which would be a common cause, the test failure will be repeated in the future.

Process Understanding: Tools and Techniques

A process diagram is a tool for defining the process which is a major focus for TQM activity. An initial step in any TQM activity should be to define and understand the organization process and the individual processes to accomplish the work. Each organization, function, and person should define their specific process(es) and understand how the process satisfies internal and external customers' needs and expectations. Each process is a customer of the preceding process, and each process has a customer for their process. Everyone must constantly strive to improve their process both as a customer and for a customer.

A process diagram uses symbols and words to describe the process. It provides an indication of improvement opportunities, non-value-added tasks, and where simplification of a process is possible. A process diagram identifies graphically the interrelationships of the process to show the roles and relationships between processes. It shows which elements impact process performance. Finally, it indicates where the process should be measured.

There are three types of process diagrams:

- Top-level process diagram
- Top-down process diagram
- Detailed process diagram

Top-level process diagram

A top-level process diagram is a picture of the entire process. This type of process diagram shows the input(s), the process, and the output(s) of a process. The top-level process diagram should focus on the satisfaction of customers' needs and expectations. The expected results must be determined, and the process must be measured to determine if it is achieving results. Figure 8.5 shows a top-level process diagram of a system development process.

Top-level process diagram steps

1. Define the specific outcome of the process focusing on the customer.

2. State the process in terms of what work has to be done to meet the outcome.

3. Determine the input(s) required to satisfy the customer.

4. Measure the process. This measurement is usually the difference between inputs and outputs.

5. Analyze how the process is performing at the top level.

6. Use improvement methodology within the continuous improvement cycle.

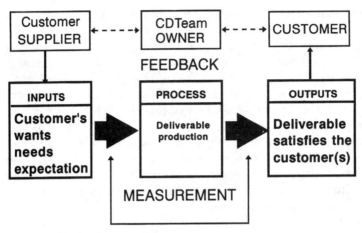

Figure 8.5 Top-level process diagram.

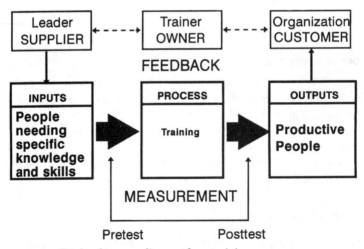

Figure 8.6 Top-level process diagram for a training process.

Top-level process diagram example

As we analyzed our system development process, we discovered that the customer expects the product to be repaired in minimum time. This quick response requires us to have trained technicians to service our product. Looking at Fig. 8.6, the top-level process diagram, we see the training process. The customer expects skilled people as the outcome of the process. Entering the process are unskilled people. The training process takes the unskilled people and transforms them into skilled people. Before the people enter training, a pretest is administered to determine entry-level skills. At the completion of training, a posttest is given to determine the overall effectiveness of training. This test is the measurement of the process. The training process is then continually checked and improved through use of the continuous improvement cycle.

Top-down process diagram

A top-down process diagram is a chart of the major steps and substeps in the process. Through examination of the major steps, the opportunities for improvement are focused on the essential steps in the process.

Top-down process diagram steps

The steps in charting a top-down process diagram are

1. List the major steps in the process. Keep it to no more than seven steps.

2. List the major substeps. Keep it to no more than seven steps.

Top-down process flow diagram example

Still using the training example, the top-down process flow diagram would list the major processes in the top-level process. When working with a team, each team member completes a process worksheet as shown in Fig. 8.7. Then the team develops the top-down process flow diagram as shown in Fig. 8.8. In the example, the major processes are analysis, design, development, implementation, and evaluation. The subprocesses under each major process are the next items listed. For instance, under the major process "analysis" the following subprocesses would be written: (1) training analysis, (2) job/process analysis, (3) task analysis. Under "design," the subprocesses are (1) course objective, (2) lessons, (3) terminal objectives for lessons, (4) enabling objectives for each terminal objective, (5) trainee measurements, and (6) training plan. Under "development," the subprocesses are (1) lesson plan, (2) training materials, and (3) training production. Under "implementation," the subprocesses are (1) prepare for presentation, (2) present the training, and (3) training management. Under "evaluation," the subprocesses are (1) plan evaluations, (2) produce evaluation instruments, (3) conduct evaluations, (4) analyze results, and (5) continuous improvement.

MAJOR PROCESSES	ANALYSIS	DESIGN	DEVELOPMENT	IMPLEMENTATION	EVALUATION	
SUBPROCESS 1	Performance analysis	Objectives	Lesson plan	Presentation techniques	Evaluation plan	
SUBPROCESS 2	Process Analysis	Lessons	Training materials development	Presentation preparation	Evaluation Instruments	
SUBPROCESS 3	Task analysis	Measurements	Production of Training Materials	Presentation delivery	Conduct Analysis	
SUBPROCESS 4		Course outline		Training administration	Analyze results of evaluation	
SUBPROCESS 5		Training plan				
SUBPROCESS 6						

Figure 8.7   Top-down process diagram worksheet.

Detailed process diagram

A detailed process diagram is a flowchart consisting of symbols and words that completely describe a process. This type of diagram provides information indicating improvement opportunities, identifying areas for data analysis, determining which elements impact process performance, and documenting and standardizing the process. It is helpful in the identification of non-value-added tasks and areas for simplification. Further, complex activities and unnecessary loops are visualized. This type of process diagram is useful for training, documentation, and explaining the process to others.

Before deciding to do a detailed process diagram, decide on the specific detail and boundaries of the process diagram. Detailed process diagrams are time-consuming. Therefore, specific boundaries are important to ensure progress on achieving improvements.

Detailed process diagram basic symbols

There are many detailed process diagram symbols. Today, there are many computer programs such as Visio and ABC Flowcharter that have numerous process diagram symbols. If you need a high level of detail, these programs are very useful. To keep it simple, four basic symbols are recommended. These symbols, as shown in Fig. 8.9, are

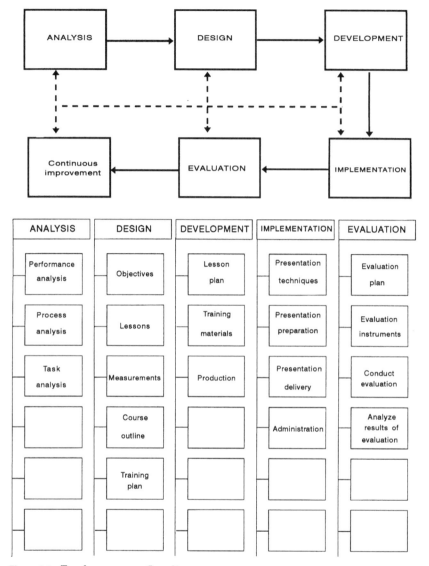

Figure 8.8   Top-down process flow diagram.

enough for most process diagramming needs. The four basic detailed
process diagram symbols are as follows:

1. Start or end is shown by squares with rounded sides.

2. Action statements are written as squares.

3. Decision statements are written as diamonds. A decision state-

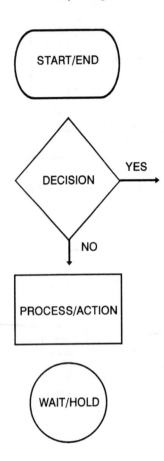

Figure 8.9 Process diagram basic symbols.

ment asks a yes or no question. If the answer is yes, the path labeled "yes" is followed; otherwise the other path is followed.

4. Wait/hold is illustrated by circles.

Detailed process diagram steps

The steps in completing a detailed process diagram are

1. Decide on specific detail and boundaries of the detailed process diagram.

2. List all the steps required in the process within the boundaries.

3. Construct a process flow diagram.

4. Determine times and cost of each activity.

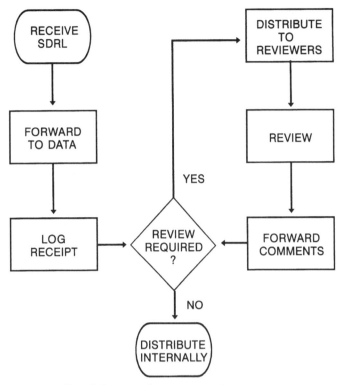

Figure 8.10   Detailed process diagram example.

Detailed process diagram example

The detailed process diagram example is shown in Fig. 8.10. It consists of the following activities:

1. A subcontractor data requirements list (SDRL) is received by materials. (input)

2. Materials forwards the SDRL to data management. (action)

3. Data management logs the SDRL. (action)

4. Data management determines if an internal review is required before distribution. (decision)

5. If an internal review is not required, the SDRL is distributed. (no)

6. If an internal review is required, data management forwards the SDRL to reviewer(s). (yes)

7. The reviewer(s) review the SDRL and make comments. (action)

8. The reviewer(s) forward the comments to data management. (action)

9. Data management determines if further review is required. (decision)

10. If no further review is required, the SDRL is distributed. (output)

Input/output analysis

Input/output analysis is a technique for identifying interdependency problems. This identification is done by defining the process and listing inputs and outputs. Once the inputs and outputs are determined, the relationship of inputs to outputs is analyzed along with the roles of the organization. The top section of Fig. 8.11 shows the input/output diagram template. The bottom section shows the inputs and outputs from the logistics process.

Input analysis

The input analysis lists all the inputs of the process. These inputs are based on the requirements of the process. Once the inputs are known, the prime owner and support responsibilities for each of the inputs is defined. This thorough analysis of all the inputs is used to match with outputs.

Output analysis

The output analysis lists all the outputs of the process. Again, the prime owner and support responsibilities are understood after communications with supplier(s), owner, and customer(s). The team especially needs to listen to the customer.

Input/output analysis steps

The steps for an input/output analysis are as follows:

1. Define the actual process.

2. List inputs and outputs of the process.

3. Determine prime (owner) and support (influencing) responsibilities.

4. Match inputs and outputs with organizations.

5. Define roles of the organizations.

6. Document on input/output analysis worksheet.

7. Use improvement methodology within the continuous improvement cycle to improve the process.

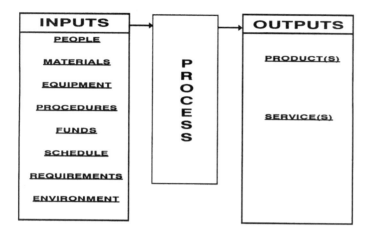

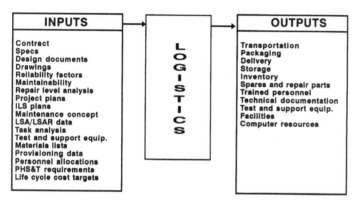

Figure 8.11 Input/output diagram template and input/output from logistics process.

Input/output analysis example

The input/output analysis uses the information from the input/output diagram to complete the input/output analysis worksheet. In the example, the input/output diagram of the logistics process shown in Fig. 8.11 is the basis for the input/output worksheet as shown in Fig. 8.12. The example shows an output of the logistics process as documentation. An input to the documentation is drawings—engineering drawings which are an input for the output documentation. The prime owner of the documentation is the technical documentation section of the logistics organization. The engineering function has the main support role of providing the drawings for the documents. The roles are interdependent; without the drawings, the document cannot be pro-

PROCESS	INPUT	OUTPUT	PRIME	SUPPORT	ROLE
Logistics	Drawings	Documentation	Tech. Pubs.	Engineering	Inter-dependent

Figure 8.12   Input/output worksheet.

duced. This relationship is shown in Fig. 8.12 on the input/output worksheet. This form can be used to look at relationships of output to input or input to output. Once this information is known, improvement opportunities can be identified and implemented through use of the continuous improvement cycle.

Supplier/customer analysis

Supplier/customer analysis is a technique that involves your suppliers in the development of your requirements and their conformance to them. In addition, it provides insight into your customers' needs and expectations and how to meet those expectations.

It is important to develop a partnership with your suppliers and a relationship with the customers you want to keep or gain. Use surveys and interviews to ensure a mutual agreement on supplier requirements and customer expectations. The supplier/customer analysis worksheet can be used to document results. You must communicate with and listen to suppliers and customers and thoroughly analyze their perceptions to continuously improve supplier performance, the process, and customer satisfaction.

Surveys and Interviews

Surveys and interviews are the major methods for getting information from suppliers and customers. Surveys involve the following:

Set a survey strategy. The team develops the survey strategy to ensure that it targets the results expected. In addition, the survey strategy focuses on providing information that is valid, consistent, and free of bias.

Use simple, concise, and clear questions. Each question must be easily understood by all respondents.

Run a pilot. This is the only method to ensure an accurate survey. The pilot should be run on a representative sample of the true population for the survey.

Use the most effective and efficient format. This method involves deciding on the best type of questions, either open or closed. Use closed questions to get specific answers. For instance, if you want to know if the customer agrees or disagrees with a particular item, ask a closed-ended question with a response scale of either agree or disagree. If you want to know how a customer feels about a particular feature, ask an open-ended question. In addition, the questions should be limited to 12 to 15 words. These types of questions are understandable. Further, ask top-priority questions first. The respondents may become tired, bored, or disinterested as the survey questions progress. Also, cluster related questions to keep the respondents' minds on the subject and to avoid confusion. In addition, provide a response scale with a thorough description. Finally, develop a scoring system which will allow for charting a detailed analysis.

Ensure room for comments. Typically, respondents' comments are valuable indicators of real information.

Yield to comments. If a respondent takes the time to make comments, the surveyor must pay particular attention to this information. Often it is worthwhile to schedule an interview with a respondent who offers comments on a survey.

Interviews involve the same considerations as surveys. In addition, the interviewer should do the following:

Instill an atmosphere of openness, honesty, and trust. This environment enables rapport to be built for the interview.

Nurture the self-esteem of the interviewee. This quality is essential, as it allows the interviewee to freely communicate.

Trust the interviewing process.   To maintain the consistency of the interview the interviewer must stay within the strategy of the interviewing process.

Empathize with the interviewee.   The interviewer needs to put her- or himself in the shoes of the interviewee during the interview.

Respond to the interviewee frequently.   Show the interviewee that you are sincere by making nonverbal gestures such as nodding approval or smiling. In addition, show that you understand by paraphrasing and summarizing often.

Wait for the interviewee to respond.   Resist the temptation to fill in gaps of silence. Silence in an interview should not be viewed as a waste of time. Allow the interviewee time to collect thoughts and formulate opinions.

Invite the interviewee to build on ideas.   Ask open-ended questions to get additional information. Use closed-ended questions to get specific answers.

Ensure that the interviewee gets feedback on results of the interview. This feedback makes the interviewee feel that he or she contributed to an outcome.

Write down all information.   The documentation of the interview is the only means to analyze results. Ensure that the information is accurate and thorough.

Supplier analysis

The supplier analysis consists of answering the following questions:

Did you survey the supplier to ensure that requirements are known?

Is there a mutual understanding of requirements?

Have you established a partnership with key suppliers?

What are suppliers' perceptions of your requirements?

Did you listen to the supplier's concerns?

Did the supplier listen to your concerns?

Were interviews conducted to determine the supplier's perceptions?

Were customer expectations translated into supplier requirements?

Is your supplier satisfying your requirements?

Customer analysis

Customer analysis seeks to answer the following questions:

Are you communicating to ensure that you are satisfying customers?

Do you understand customers' needs and expectations?

Have you conducted a survey to determine if you are satisfying your customers?

Has a thorough analysis been completed to ensure that the entire process is focused on customers' needs and expectations?

Does the owner understand process impacts on customers?

Are process outputs measured in relation to customers' expectations?

Are you satisfying mutually agreed upon customer expectations?

Have you developed a relationship with key customers?

Supplier/customer analysis steps

The supplier/customer analysis steps are

1. Identify the customers (both internal and external) of the process.
2. Determine the needs and expectations of your customers.
3. Identify the products or services you provide to meet those needs and expectations.
4. Develop measures of your output that reflect customer expectations.
5. Determine if the customer expectations have been met or not met.
6. Determine who owns or influences the product or service.
7. Identify your principal inputs (manpower, material, machine, method, environment).
8. Determine if suppliers know their requirements and their impact on your meeting customer expectations.
9. Involve your suppliers in the development of your requirements and their conformance to them.
10. Identify suppliers that are not meeting requirements.
11. Document results on supplier/customer analysis worksheet.
12. Use structured improvement methodology to improve supplier performance, the process, and customer satisfaction.

Supplier/customer analysis example

The supplier/customer analysis example continues using the engineering drawings required for documentation. First, transfer the input, output, and supplier information from the input/output worksheet to the

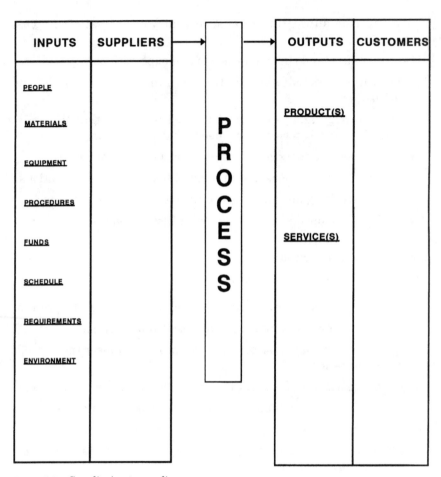

Figure 8.13 Supplier/customer diagram.

supplier/customer diagram as shown in Fig. 8.13. Second, transfer the information from the supplier/customer diagram to the supplier/customer worksheet as shown in Fig. 8.14. In the example, the input of drawings provides the data for the maintenance documentation output. The customer expects a usable document for maintenance personnel. Information for this documentation is obtained by communication with the customer. Third, a specific measurement must be determined. In this case, the customer expects the document to be 80 percent accurate. Fourth, measure performance. Fifth, determine if the document is meeting or not meeting customer expectations.

To complete the supplier side of the equation, engineering must be informed of the drawing's impact on the provision of an accurate doc-

INPUT	SUPPLIER	REQUIREMENT	MET/ NOT MET	OUTPUT	CUSTOMER	NEED/ EXPECTATION	MET/ NOT MET
Drawing	Engineering	Complete and accurate	Met	Repair Manuals	Maintenance People	100% accurate	Met

Figure 8.14   Supplier/customer worksheet.

ument to the maintenance personnel. The requirement for complete and accurate drawings must be measured to determine if engineering is meeting or not meeting requirements.

All the information from the supplier/customer worksheet is used within the improvement methodology to improve supplier performance, the process, and customer satisfaction.

Benchmarking

Benchmarking is a method of measuring your organization against those of the recognized best performers in a certain industry, organization, function, system, or process. The purpose of benchmarking is to provide a target for improving the performance of the organization. The benchmark targets improvement of the process outputs or the performing of the actual process. Benchmarking brings a focus on customer-driven project management improvement efforts by emphasizing desired outcomes. It also nurtures wholesome competition by creating the desire to be the best. Benchmarking provides a common focus to hold the organization together by measuring critical areas and analyzing these critical areas against the best. This targeting of the best reinforces continuous improvement by keeping everyone aiming at a long-term objective.

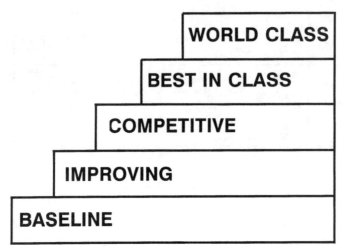

Figure 8.15   Steps toward "world class."

Figure 8.15 gives an example of steps to "world class." The leaders are considered world class. The organization starts with its current performance. This is the baseline. Through the implementation of continuous improvement, the organization moves to improving. As the organization institutionalizes continuous improvement, it progresses to competitive, best in class, and world class. With the help of benchmarking, this continuous improvement can be planned and implemented to meet the organization's specific objectives.

Four types of benchmarking

There are four methods of benchmarking: internal, competitive, functional, and generic. In each case, the type of benchmarking selected depends on the measures needed and the methods used to collect the data.

Internal benchmarking.   Internal benchmarking looks inside the organization for similar processes and units that seem to do it better.

Competitive benchmarking.   Competitive benchmarking looks at competitors and examines their processes. This type of benchmarking seeks other institutions that are performing better than the customer-driven project management organization. When these processes are found, the competitors' performance is compared to the customer-driven project management organization.

Functional benchmarking.   Functional benchmarking looks at any outside or inside activity that is functionally exact to the process under review.

Generic benchmarking.   Generic benchmarking looks at any outside or inside activity that is generically the same as the one under review.

Benchmarking considerations

Builds a target for improvement efforts

Emphasizes desired outcomes

Nurtures competitiveness

Creates the desire to be the best

Holds the organization together striving for excellence

Measures critical areas

Analyzes critical areas against the best

Reinforces continuous improvement

Keeps everyone moving toward a target

Benchmarking steps

1. Understand your organization
2. Select critical areas for benchmarking
3. Determine where to get benchmark information
4. Collect and analyze data
5. Select target benchmarks
6. Determine your performance
7. Set desired outcomes
8. Use improvement methodology to achieve desired performance

Benchmarking example

Benchmarking starts with the strategic intent of the organization performing the benchmarking process. There must be a commitment from the top leadership in the organization to pursue continuous improvement with benchmarking as a tool. The following provides an example of one cycle through the benchmarking process beginning at the top leadership of an organizational development and training (OD&T) organization.

Step 1: Understand your organization.

Purpose: Define the focus of the benchmarks.

The benchmarking process begins as a result of strategic planning. A complete understanding is needed of all the areas to meet total customer satisfaction as outlined in the beginning of this chapter. The mission outcome of strategic planning provides the focal point for benchmarks for the organization. In our example, the mission is as follows:

> Our mission is to be the leading OD&T organization for individuals and organizations seeking continuous improvement focused on total customer satisfaction by delivering continuously improving, value-added, results-oriented, customer-satisfying organizational development and training products and services.

Step 2: Select critical areas for benchmarking.

Purpose: Determine what to benchmark.

The second step involves listing the areas considered significant for success of the mission, namely the customer needs and expectations, deliverables to meet the customers' specification, and the internal processes to satisfy the requirements. Quality Function Deployment (QFD) phase 1, outlined in Chapter 9, provides an excellent tool for listening to the "voice of the customer." In Fig. 8.16 the top of the chart shows some of the areas for consideration by the OD&T organization. From this list, the organization selects the critical areas for benchmarking. The OD&T organization decides to benchmark all of the areas critical to customer satisfaction at this stage. During other stages in the TQM improvement methodology, the team may select other processes to benchmark.

The OD&T critical areas of customer satisfaction include the following:

- Personal—Ability to adapt the deliverable to specific customer needs and expectations

  Measure: Percentage of special requests met

- Responsive—Capability to meet the needs and expectations of customers

  Measure: Percentage of total customer requests met

- Obtainable—Ability to provide deliverable within customer's affordability

  Measure: Percentage of customer's loss due to cost

- Deliverable—Provide deliverable when customer needs it

Measure: Percentage of on-time delivery

- **U**seful—Deliverable provides business results

Measure: Percentage of customers reporting business results within 30 days

- **C**onvenient—Provide deliverable where the customer wants it

Measure: Percentage of requests for specific location met

- **T**imely—Ability to provide OD&T solution at the time needed by the customer

Measure: Percentage of times deliverable is just in time for customer's results

- **S**atisfaction—Ability to totally satisfy the customer

Measure: Number of customer complaints

OD&T evaluation rating on 5-point scale

Number of customers requesting refunds from money-back guarantee

Step 3: Determine where to get benchmark information.

Purpose: Find sources of benchmarking data.

Since the benchmarking information becomes the target, getting the right benchmarking information is the most important aspect of benchmarking. The sources of information for process performance measurements are numerous. The only real source of a benchmark for performing an actual process is the process performing organization. Some sources of benchmarking data include the following:

- Computer databases
- Industry publications
- Professional society publications
- Company annual reports and publications
- Conferences, seminars, and workshops
- Other organizations within the same organization
- Consultants
- Site visits

Major Processes

Customer Requirement	Deliverables	Processes
- Personal	- OD	- Training
- Responsive	- Training	- OD
- Obtainable	- Interventions	- Administration
- Deliverable	- Consulting	- Finance
- Useful	- Coaching	- Scheduling
- Convenient	- Facilitating	- Information systems
- Timely	- Assessing	- Support
- Satisfaction	- Support services	

Critical Areas for Success

	Measures of Success	Current Performance	Benchmark Target
Personal	% of special requests met	72%	98%
Responsive	% of total customer requests met	76%	100%
Obtainable	% of customers lost due to cost	25%	10%
Deliverable	% on time delivery	85%	100%
Useful	% of customers reporting business results	83%	100%
Convenient	% of requests for special location met	82%	98%
Timely	% of just in time deliverables	89%	100%
Satisfaction	Number of complaints	5 per month	5 per year
	Evaluation rating on 5 point scale	4.2	4.9
	Number of customer refunds	10 per year	0

Goals

	Year 1	Year 2	Year 3	Year 4
Personal	80%	90%	98%	100%
Responsive	80%	90%	100%	100%
Obtainable	20%	15%	10%	0%
Deliverable	90%	95%	100%	100%
Useful	90%	95%	100%	100%
Convenient	95%	95%	98%	100%
Timely	95%	98%	100%	100%
Satisfaction	1 per month	8 per year	5 per year	0
	4.5	4.8	4.9	5
	5 per year	1 per year	0	0

Figure 8.16 Benchmarking chart sample: (top) major processes; (middle) critical areas; (bottom) goals.

For example, the organizational development and training organization selected the following sources of information:

- American Society of Training and Development
- *Training Magazine*
- Annual reports of leading organization and training organizations

Step 4: Collect and analyze data.

Purpose: Get the right and accurate data.

Once the sources of information are determined, the data is collected and analyzed. First, the data collected for benchmarking must be the right information. It needs to be the information that fits the organization's requirement for benchmarking. It must truly reflect the data for a leader. Second, the information must be analyzed to ensure it is accurate. It is necessary to verify and validate any process performance measurement information as applicable to your specific operation before using it as a benchmark in the organization. In addition, it is imperative to verify and validate process operations by the direct observation of the process.

Step 5: Select target benchmarks.

Purpose: Establish long-range focus.

During this step, the OD&T organization selects the target benchmarks to meet its mission. The middle section of Fig. 8.15 shows the critical areas for success, and sample target benchmarks are listed in the last column.

Step 6: Determine your performance.

Purpose: Know how the organization is currently performing.

During this step the selected critical areas are measured. To do so, metrics must be developed. The metrics development process is described later in this chapter. The middle section of Fig. 8.16 lists current performance in the center column.

Step 7: Set desired outcomes.

Purpose: Establish short- and long-term goals to achieve targets.

The benchmarked target may take several years to achieve, depending on the current performance and capability of the organization. The organization establishes a plan to achieve the benchmark. The plan may be set up to achieve the benchmark in stages like the steps toward world class shown in Fig. 8.15. The bottom section of Fig. 8.16 shows typical goals to reach the target. In year 1, the OD&T organization aims for improvement over their current performance. In year 2, they seek to be competitive. In year 3, they target best-in-class performance goals. In year 4, the OD&T organization strives for world-class status.

Step 8: Use improvement methodology to achieve desired performance.

Purpose: Use a systematic approach to reach benchmark.

Finally, the benchmarking and improvement process is continuous. The organization must establish a continuous improvement system to achieve the target.

Metrics

Metrics are measurements made over time that communicate vital information about the quality of a process, activity, or resource. They reflect meaningful measures that target continuous improvement actions. This type of measurement is differentiated from plain measurement by its specific focus on total customer satisfaction while supporting the organization. Metrics must be customer-oriented and communicate a state of health; show where we are now in relation to where we want to go over time; and assess all critical processes and activities leading to success.

Metrics considerations

Customer-oriented metrics foster understanding and help develop a trusting relationship. The customer agrees that this type of measurement is an accurate indication of customer satisfaction. Good metrics always focus results toward an improvement action. Metrics must distinguish between acceptable and unacceptable actions; target long-term improvement; indicate a trend over time while being timely; be unambiguously defined with a specific link to the organization's objectives; and be easy and economical to gather.

The following are important metrics considerations:

Meaningful to the customer (internal/external)

Establishes appropriate action

Tells how well the organization is performing

Repeatable over a period of time

Indicates a trend

Clear operational definition

Simple to collect

Metrics are not...

- **Charts.** Charts may graphically display the results of metrics, but the chart itself is not a metric.

- **Schedules.** Some forms of schedules can lead to good metrics, but usually schedules do not provide information that by itself will lead to improvement.

- **Goals, objectives, strategies, plans, missions, or guiding principles.** Most of these can be measured, but metrics are not by themselves the end. They are a means to an end.

- **Counts of activity.** Counts of activity can result in metrics, but just because you have a measure does not necessarily drive appropriate action.

- **Snapshots or one-time status measures.** These show little trend information. A comparison of status over time can be metrics but tends to be very top level and does not provide a "real" understanding for specific action.

The metrics package

The metrics package consists of the following three basic elements: the operational definition, the actual measurement, and metric presentation. The operational definition is the precise explanation of the process being measured. The measurement involves the collection, sorting, and translation of the data from the process into understandable and useful information. The metric presentation provides the communication link to the customer (internal/external). Figure 8.17 shows the parts of metrics package.

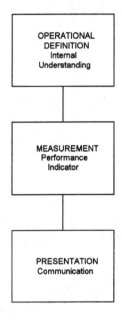

Figure 8.17   Metrics package.

Developing metrics action process

1. Define the purpose of the metric.
2. Develop the operational definition.
3. Determine if measurements are already available
4. Generate new measurements if required.
5. Evaluate the validity of the metrics.
6. Institute and baseline the metrics.
7. Measure progress against the baseline.
8. Prepare the metric presentation.
9. Use the metric for continuous improvement action.

Metrics example

The following provides a step-by-step example of metrics development for the customer-driven project team to improve its effectiveness.

Step 1: Define the purpose of the metric.

Purpose: Improve customer-driven team's effectiveness.

The purpose of the metric must be focused on a specific process geared to meeting the customer's needs and expectations. In this case, the process is project performance achieving total customer satisfaction.

Step 2: Develop operational definition.

Operational definition: The customer-driven project team will perform a monthly project performance review at the second team meeting of each month using an approved "customer satisfaction critique."

The operational definition provides the who, what, when, where, and how of the metric. In this example, the operational definition includes the following:

- *Who:* Customer-driven project team
- *What:* Perform "customer satisfaction critique"
- *When:* Monthly, second meeting of each month
- *Where:* Team meeting
- *How:* Presentation using approved "customer satisfaction critique"

Step 3: Determine if measurements are available.

Current measurements do not meet criteria for metric.

Although there are many existing measurements for project performance, they do not focus on customer satisfaction. They only target cost, schedule, and technical specifications. In this case, new measurements need to be generated. If at all possible, use existing measurements. Do not invent an unnecessary measurement.

Step 4: Generate new measurements if required.

New measurements include

- Deliverable satisfaction
- Project execution: schedule, cost, and technical
- Customer relationship
- Other working relationships: supplier, other CDTs, and so forth
- Team leader performance: team meetings, keeping focused, problem solving, decision making, team building, managing conflict

These metrics focus on the performance measurement of the key processes for project performance. If we measure these items that provide an indication of customer satisfaction, the customer-driven project team's performance will improve along with its processes.

Step 5: Evaluate the validity of the metric.

Rating of metric:

- Approved by customer-driven teams? Yes
- Linked to organizational objectives? Yes
- Is simple and clear? Yes
- Can display a trend over time? Yes
- Is easy to collect data? Yes
- Can be used to drive process improvement? Yes

Step 6: Institute and baseline the metric.

Measurement tool: Customer satisfaction critique

During this step, the measurement tool is selected and initial data are collected. For this example, a customer satisfaction critique was selected for collecting the data. Figure 8.18 shows a sample customer satisfaction critique. The results of the critique will be displayed on a line chart. Figure 8.19 shows the initial customer satisfaction charts.

Step 7: Measure the process against the baseline.

Each month the customer satisfaction critique is distributed, collected, and charted to examine trends over time.

Step 8: Prepare metric presentation.

Each month the metric presentation is prepared. The metric presentation consists of the metric description and the metric graphic. Figure 8.20 displays the metric description for the example. The metric description consists of the metric operational definition, the measurement method, the desired outcome, the linkage to organizational objectives, and the process owner. An example of a metric graphic is shown in Figure 8.21 for deliverable satisfaction. Besides the graphic data, a "stoplight" chart is integrated with the metric graphic to highlight status. A stoplight chart simply uses green, yellow, and red to indicate a particular assessment. For instance, green indicates no problems. Yellow means that some potential problems exist. Red indicates that a problem exists.

Customer Satisfaction Critique

Instructions	Please rate the customer-driven team based on the 5 point scales below. Circle the number on each scale that best states your opinion at this time.

1. Deliverable satisfaction

Is the customer satisfied with the deliverable to this point?

Very
dissatisfied   1   2   3   4   5   Very
satisfied

2. Relationships

Is the customer satisfied with the current customer relationship?

Very
dissatisfied   1   2   3   4   5   Very
satisfied

Is the customer satisfied with the team's relationships with suppliers, other teams, other leaders?

Very
dissatisfied   1   2   3   4   5   Very
satisfied

3. Project execution

Is the customer satisfied with schedule, cost, and technical performance?

Very
dissatisfied   1   2   3   4   5   Very
satisfied

4. Team leader performance

Is the customer satisfied with the team leader's performance?

Very
dissatisfied   1   2   3   4   5   Very
satisfied

Figure 8.18 Customer satisfaction critique.

DELIVERABLE SATISFACTION

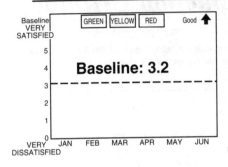

PROJECT EXECUTION

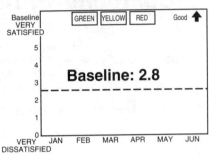

CUSTOMER RELATIONSHIP

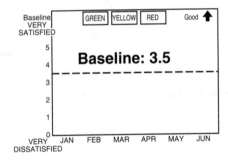

OTHER RELATIONSHIPS

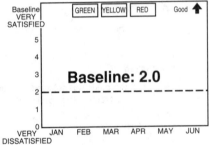

TEAM LEADER PERFORMANCE

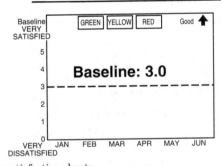

Figure 8.19 Customer satisfaction charts.

METRIC TITLE:

IMPROVE CDT's EFFECTIVENESS

Operational Definition:

The customer-driven project team will perform a monthly project performance review at the second team meeting of each month using the results of the approved customer satisfaction critique.

Measurement Method:

The measurement consists of a customer satisfaction index complied as an average from the approved customer satisfaction critique. The customer satisfaction index indicates the defined key customer's satisfaction with deliverable satisfaction, project execution, customer relationships, relationships with other key interfaces, and team leader performance. The customer satisfaction index uses a closed scale from 1(very dissatisfied) to 5(very satisfied).

Desired Outcome:

Improve customer satisfaction rating in all five areas.

Linkage to Organizational Objective:

Meets total customer satisfaction strategy.

Process Owner:

Customer-driven project team

Figure 8.20 Metric description example.

DELIVERABLE SATISFACTION

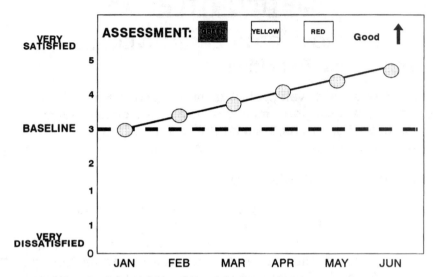

Figure 8.21   Metric graphic—deliverable satisfaction.

Step 9: Use the metric for continuous improvement action.

Based on the results of the metric analysis, the team leader initiates process improvement actions. These actions must be focused on continuous improvement.

Selection Tools and Techniques

Selection techniques are used to help clarify assumptions and focus on consensus. The first step to selection involves choosing a technique from voting, the selection matrix, and the selection grid. Second, the selection criteria must be established by the team. Third, the team must have a list of items to select. The list consists of issues, problems, or opportunities when an item needs to be selected for the team to work on or alternatives when a solution needs to be selected. Fourth, the team must communicate to get an understanding of each item for selection. This communication involves discussing trade-offs of ideas with everyone involved. Fifth, the team needs to obtain consensus on the selection. This selection must meet the needs of the team for a problem to solve, an opportunity to improve, or a solution to implement. Finally, the team nurtures the decision by communicating the benefits with key people. The selection process can be summarized as follows:

Select a technique.

Establish criteria.

List issues, problems, opportunities, or alternatives.

Evaluate the list using the criteria.

Communicate until the team reaches an understanding.

Trade off ideas.

Involve everyone affected.

Obtain consensus.

Nurture the decision.

Selection techniques steps

The steps of selection techniques are as follows:

1. List issues, problems, opportunities, alternatives.
2. Determine criteria.
3. Select method.
   - Vote
   - Matrix
   - Grid
4. Make a decision.

Selection techniques lists

The list of issues, problems, and opportunities is usually the result of a brainstorming session, process diagrams, input/output analysis, and customer/supplier analysis. Alternatives are generated from brainstorming sessions or force-field analysis.

Selection techniques criteria

The team must determine realistic criteria to make the selection. There are many factors that could be considered for selection criteria. Although the team can select any criteria, the criteria below provide examples for consideration:

Cost

Resources

Importance

Time

Effect

Risk

Integration with organization's objectives

Authority

Sometimes cost may not be a factor in the selection, but usually it is a primary consideration. The specific cost factor should be stated in the criteria. For example, a total cost of less than $800 may be a selection criterion.

Resources involve many individual factors, including capital, labor, equipment, and natural resources; they usually are a major consideration in any selection. Again, the specific criteria must be determined by the team. Many times the selection must be made based on no increase in resources of any kind.

The importance of the project to the organization or team may be another criterion. Normally, a team does not want to waste time on unimportant projects. The team should always strive to focus on the vital few versus the trivial many. Again, there could be much debate on the importance of any single item. The specific criteria may have to be stated to avoid confusion.

Time is always a necessary criterion in any selection. Without a time criterion, almost anything is possible. Furthermore, a time criterion targets a selection geared to short-term or long-term results.

Risk is a consideration many organizations use as a criterion. The amount of risk the team or the organization is willing to assume should be stated for selection criteria.

Another selection criterion often overlooked is integration with the organization's objectives. Sometimes a solution is selected beyond the scope of the organization's focus.

The final criterion offered is authority, which should always be one of the selection criteria. Frequently, teams select a problem or a solution beyond their authority. Then they become frustrated with efforts to get others to implement their solution. The team needs the wisdom to work on the issues and solutions they own.

Selection techniques: voting

Voting is a selection technique used to determine majority opinion. This technique may be useful in narrowing a list of problems, opportunities, or alternatives. Since this method often leads to a win-lose situation, it is not recommended as a final decision-making technique for selecting an issue, problem, opportunity, or solution. The objective of all selection techniques is to focus on reaching consensus on a win-win solution. Again, this is just a preliminary step to limit the number of items to get a consensus. There are three primary voting techniques: (1) simply having people raise their hand to indicate a vote so the item with the most votes can be selected (this method is effective for voting on a small number of items), (2) rank-order voting, and (3) multivoting (the latter two are more effective for voting on a large number of items).

Rank-order voting

Rank-order voting is a quick method for ranking a list of issues, problems, opportunities, and alternatives to determine the top priorities.

Rank-order voting steps

The steps in rank-order voting are as follows:

1. Generate a list of items requiring a decision.
2. Combine similar items.
3. Number items.
4. Have each member rate each item on a scale of 1 to 5, with 5 being the high number.
5. Total the points for each item.
6. Rank the items from highest to lowest based on total points.
7. Reach consensus on the top priority.

Rank-order voting example

The following is an example of rank-order method of selection. First, in a brainstorming session concerning people issues a list of items for selection as opportunities was generated at a specific organization. Second, similar items were combined. Third, items were numbered. Fourth, each member rated each item. Fifth, the total points were listed.

1. Training	24
2. Morale	17
3. Location of personnel	18
4. Not interested in someone else's responsibility	21
5. Staffing problems	23
6. Bad attitudes	16
7. Tendency to separate people	13
8. People's dissatisfaction	18
9. Poor workmanship	20
10.Lack of right skills	15
11.Lack of indirect resources	21
12.Not right people on shift	16
13.Lack of team spirit	20
14.No ownership	23
15.Understaffed	22
16.No accountability	21
17.Restricted advancement	18
18.Aging work force	18
19.Lack of pride in work	17
20.Too many leaders; not enough workers	25

Sixth, the list was rearranged highest to lowest:

1. Too many leaders; not enough workers	25
2. Training	24
3. Staffing problems	23
4. No ownership	23
5. Understaffed	22
6. Lack of indirect resources	21
7. No accountability	21
8. Not interested in someone else's responsibility	21
9. Poor workmanship	20
10. Lack of team spirit	20
11. People's dissatisfaction	18
12. Restricted advancement	18
13. Aging work force	18
14. Morale	17
15. Lack of pride in work	17
16. Bad attitudes	16
17. Not right people on shift	16
18. Lack of right skills	15
19. Tendency to separate people	13
20. Location of personnel	8

Seventh, the team reached a consensus on the selection. In this particular example the highest number of points was "too many leaders; not enough workers." However, the team decided through consensus to select "training." This was the issue the team could impact.

Multivoting

Multivoting is a technique used to reduce a large list of issues, problems, opportunities, or alternatives to a smaller number of items.

Typically, a team during the understanding stage has a large list of concerns with the process. Then this list must be reduced to focus the team on a consensus of the most critical issues to work.

Steps in multivoting are as follows:

1. Generate a list of items requiring a decision.

2. Combine similar items.

3. Number items.

4. Have each member select a number of items from the list as follows: Each person should write the number of the items on separate paper. He or she should then select a specific number of items from that list. For example, if the total number of items on the list is 50, they are numbered 1 to 50. Each person then may be asked to select the top 20 items from the list of 50.

5. Total the number of votes for each item.

6. Create a new list of items of the top 20 vote getters. Number the items 1 through 20.

7. Have each member select a lower number of items from the list. For example, from the list of 20 items, each member may be asked to select the top 5 items.

8. Reach consensus on the top-priority items.

Multivoting example

Figure 8.22 shows an example of multivoting. In this example, one team was looking into the issues related to methods in the organization. First, a list of 20 items were brainstormed. Then this list was pared down to the top four items. Finally, the team reached consensus on "too many signatures and approval" as the issue to focus their improvement efforts.

Selection Matrix

A selection matrix is a technique for rating issues, problems, opportunities, or alternatives based on specific criteria. The issues, problems, opportunities, or alternatives are listed on the left side of the matrix. The team selects the criteria to be considered in evaluating the alternatives. This criteria is placed along the top of the matrix. Then, the members individually rate the issues, problems, opportunities, or alternatives. The selection matrix should always be completed individually first. Next, the group ratings are determined. The results of the selection matrix do not set the actual decision. The purpose of the ma-

Issues relating to methods in the organization include:
1. Unclear administrative procedures
2. Too much paperwork
3. Lack of configuration management
4. Too many signatures and approvals
5. Numerous engineering changes
6. Outdated methods
7. Policies and procedures overkill
8. Lack of understanding customer requirements
9. Delayed work authorizations
10. No scheduling
11. Poor planning
12. Firefighting
13. No data retrieval system
14. Inadequate design reviews
15. Lack of coordination
16. Design intent versus manufacturing producibility
17. We have always done it that way
18. Distribution
19. Information availability
20. Turf protection

Select the four most critical issues. Your selections should consider the following criteria: (1) Would require no resources to solve. (2) The team can solve the issue. (3) The issue can be solved in 30 days.

1 Unclear administrative procedures

2 Too many signatures and approvals

3 Outdated methods

4 Lack of understanding customer requirements

<u>Selection is:</u>

Too many signatures and approvals

Figure 8.22 Multivoting example.

trix is to lead to a more focused discussion of each item. This process helps to clarify assumptions and focus consensus.

Selection matrix steps

The selection matrix steps are as follows:

1. The issues, problems, opportunities, or alternatives are listed on the left side of the matrix.

2. The team selects the criteria to be considered in evaluating the problems, opportunities, or alternatives.

3. The criteria is listed at the top of the matrix.

4. The members individually rate the problems, opportunities, or alternatives.

5. Each team member's highest total point item is tabulated.

6. Discussion of the issues, problems, opportunities, or alternatives ensues.

7. The group tries to reach a consensus.

Selection matrix example

The selection matrix example is shown in Fig. 8.23. In the example, the organization needs to select a method of training which ensures that test technicians can perform a certain test. The opportunities—formal classroom, on-the-job, combination of both, or none—are listed down the right side of the example. The criteria are listed across the top. For this example, the following criteria were selected: effect, cost, and time to implement. Each opportunity is rated against the criteria individually by each team member. The team member that completed the selection matrix in Fig. 8.23 rated the combination of both formal and on-the-job training as the highest solution. This solution received 22 points. The issue, problem, opportunity or solution is then rated by the total number of team members giving it the highest number of total points. For instance, the team considering the method of training consisted of eight members. The results of the eight team members' selection matrixes are as follows:

Formal classroom	3 members rated this highest
On-the-job training	2 members rated this highest
Combination of both	5 members rated this highest
None	0 members rated this highest

Issue Opportunity Problem Alternative	EFFECT High- 10 Low - 1	COST High- 10 Low - 1	TIME High- 10 Low - 1	OTHER High- 10 Low - 1	TOTAL
Formal classroom	8	6	3		17
On-the-Job Training	4	8	7		19
Combination of both	9	7	6		22
None	1	7	10		18

Team Total

ITEM	HIGHEST ITEM	TOTAL
Formal classroom	I I I	3
On-the-job training	I I	2
Combination of both	I I I I I	5
None		0

Figure 8.23 Selection matrix example.

Figure 8.23 on the bottom shows a data collection chart for team member totals. Once the highest rating of all team members is known, the group focuses on consensus. The final selection is based on the understanding and discussion of the items on the selection matrix.

Selection grid

A selection grid compares each issue, problem, opportunity, or alternative against others using the criteria.

Selection grid steps

1. List problems, opportunities, or alternatives.

2. Determine criteria.

3. Compare each pair of issues, problems, opportunities, or alternatives against others using criteria.

4. Try to reach a consensus as a group.

Selection grid example

In this example, we will continue the example of a team looking into the issues related to methods in the organization. Remember, the team used multivoting to narrow the list of issues to the top four issues. The top four issues were the following:

1. Unclear administrative procedures

2. Too many signatures and approvals

3. Outdated methods

4. Lack of understanding of customer requirements

In this example, the selection grid is used to focus on consensus. The selection grid is shown in Fig. 8.24. Again, the selection grid should be completed individually. First, each choice is given a number as shown above. Then each choice is compared against the other choices. The selection is made by circling the preferred choice. For instance, block 1, item 1, "unclear administrative procedures" is compared to item 2, "too many signatures and approvals." This team member selected item 2, "too many signatures and approvals," by circling item 2 on the selection grid. This procedure is repeated in each of the blocks. Once the selection grid is completed, the choices are totaled. The bottom left portion of Fig. 8.22 shows a data collection chart for totaling the individual selections. The individual data collection chart shows item 4, "lack of understanding of customer requirements" as the highest selection, the ini-

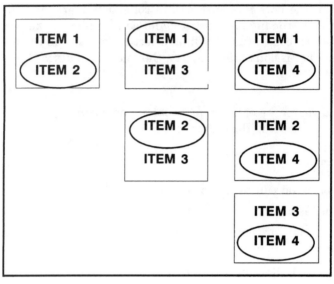

Individual Total

ITEM	CHOICE	TOTAL
ITEM 1	I	1
ITEM 2	I I	2
ITEM 3		0
ITEM 4	I I I	3

Team Total

ITEM	HIGHEST ITEM	TOTAL
ITEM 1	I I	2
ITEM 2	I I	2
ITEM 3	I	1
ITEM 4	I I I I I	5

Figure 8.24   Selection grid example.

tial choice of the team member(s). The selection grid can be completed either individually or by the entire team. If it is completed individually, the team's selections must be totaled as with the selection matrix. The bottom right of Fig. 8.24 shows the team's data collection chart. The team's selection was also item 4. Again, from this process, assumptions are clarified and a consensus can be reached. Remember, the ultimate selection may differ from the result on the selection grid.

Decision Making

Decision making is the process of making the selection. In making a decision, the impact and support of the outcome should be considered. A group will be more committed to success if the decision is reached by consensus. Therefore, consensus should be reached on the selection of an issue, problem, or opportunity to work as a team. It is also necessary when decisions are made as to solutions to implement. Consensus decision making targets a win-win outcome. Decisions reached by any other method than consensus results in a win-lose situation. A win-lose decision results in not having total commitment and support for the selection.

Although consensus is the recommended method for team decision making, other methods of decision making also exist. These types of decision making may be appropriate at times other than those mentioned above as essential consensus decision-making times. In many instances, time constraints, insignificance, or other considerations make consensus decision making unrealistic. In certain situations, the team determines an alternate method of decision making from the methods below:

- *Decision by majority.* This is a decision by more than half of the representatives.

- *Decision by leader.* In some cases, the leader makes the decision.

- *Decision by management.* Management sometimes must make the decision.

Consensus

Consensus means that everyone in the group accepts and supports the decision. It does not mean that everyone wants the selection but that everyone on the team agrees to the decision. Consensus equals commitment. It can only be reached by open and fair communication among all team members. Consensus is critical when one is selecting a process to improve, problem to solve, mission to accomplish, opportunity to pursue, recommendation, and a solution. Consensus requires understanding and discussion and it is arrived at by understanding the process, mission, problem, and all the possible alternatives. Further, discussion of all the possible driving and restraining forces, causes and effects, and process interactions from all the viewpoints of the group is absolutely essential. Once understanding and discussion takes place, the group can proceed with the process of arriving at a consensus.

Consensus involves communicating, especially listening to others' points of view. It means opening the team members' minds to new ideas. It requires a nurturing of the feelings and ideas of all team members. It allows a sharing of information and it encourages participation. Consensus nurtures discussion in that it does not encourage voting or agreeing too quickly. It fosters the support ideas that are best for everyone with the understanding that differences are a strength. Consensus decision making seeks a win-win solution.

Consensus decision making is detailed in Chapter 6.

Decision by the majority or leader

In many cases during routine group activities, decisions may be made by the majority of the membership or by the leader. Decisions made by the majority or the leader are usually reserved for relatively minor aspects of group activities. For example, the decision to use a specific tool or technique may be determined by the majority or the leader. In all cases the particular method of decision making should always be determined by the group.

Decision by management

Management may make some decisions. Decisions by management are necessary in many cases. In a customer-driven project management environment, management should always be aware of the impact of decisions on maintaining the environment. As with consensus, decisions by management, accompanied by understanding and discussion, frequently lead to support of even unpopular decisions.

Process Analysis

Process analysis is a tool used to improve a process by eliminating non-value-added activities and waits and/or by simplifying the process. Process analysis focuses on specific defined outcomes. These desired results usually aim at time and/or cost reduction. Process analysis is an extremely useful means for getting the output of the process to the customer as quickly as possible at the lowest possible cost. The major goals of process analysis are elimination or reduction of high costs; non-value-added processes, activities, and tasks; and the waits between processes.

High-cost areas are usually a primary area of focus. To determine processes of excessive costs, cost figures are added to the process diagram; then the team does a complete process analysis on each of the high-cost areas.

In addition to high-cost areas, money is lost by many organizations in performing non-value-added processes, which are another target for process analysis. The value and nonvalue of a particular process, activity, or task is a judgment based on facts within a specific environment. Each process, activity, and task deserves a thorough analysis to determine its value. The customer-driven team conducts the process analysis with inputs from all people affected by the process under review. As part of the process analysis, the customer-driven team conducts a risk assessment to reduce the probability of eliminating an essential task. Processes, activities, and tasks are not eliminated without the concurrence of the process owner and consensus of everyone affected by the process improvement.

Once non-value-added tasks are evaluated for possible savings, the team can focus on reducing or eliminating waits. Many hours are wasted between the performance of processes. This downtime affects the organization's ability to rapidly respond to customers. Rapid response time is a major differentiator in today's economic times. By concentrating on improving the waits between processes, the team can make a considerable impact on its cycle time.

During process analysis, the team first challenges the following:

- Excessive costs

- Inordinate waits

- Bureaucratic procedures

- Duplicate efforts

- Inspection or overseer operations

- Layers of approval

Once the above is examined, process simplification becomes the next step. This step involves probing the high-cost and high-time processes for simple, innovative, and creative improvements for accomplishing the process. Can the process be combined with another activity? Can the process be done less frequently? Can the process be automated to be accomplished more quickly? Can the process be done another way? These initial actions achieve quick results at little or no cost. This level of process analysis aims at process improvement to achieve increased financial performance, improved operating procedures, and greater customer satisfaction.

During this step, the team challenges the following:

- Complexity

- Unnecessary loops

- Frequency

- Methodology

Process analysis steps

1. Construct a process diagram (top-down or detailed).

2. Ensure that waits between processes/activities are identified.

3. Determine time and cost of each process/activity and time of waits.

4. Reduce or eliminate waits.

5. Select critical activities (high time or cost).

6. Eliminate non-value-added processes/activities.

7. Eliminate parts of the process.

8. Simplify value-added processes/activities. During this step, look to combine processes/activities, change the amount of time or frequency, do processes/activities in parallel with another process, and use another method to do process.

9. Use continuous improvement methodology to further improve the process.

Process analysis example

The example uses the detailed subcontractor data requirements list (SDRL) process flow diagram. Figure 8.25 shows the process flow diagram with times and Fig. 8.26 shows the process flow diagram with costs. By examining the process flow diagram, the team decides that the waits are extremely excessive. This would be their first area of focus for improvement. The team builds a timeline of activities and waits (see the bottom of Fig. 8.25). The process with an internal review is shown to take 41 days, including 23 days of waits and 18 days of actual work. The team decides to aim at reducing the waits by 50 percent. The total time would then be reduced to 28.5 days, with 8.5 days of waits. Next the team targets high-time and high-cost processes. In the example, the high-time and high-cost process is the review process, as shown in Figs. 8.25 and 8.26. The team would next do a detailed process diagram on just the review process. This process diagram would be used to eliminate non-value-added activities and simplify value-added activities.

Work-Flow Analysis

A work-flow analysis looks at a picture of how the work actually flows through an organization or facility. Like the process analysis which focuses on eliminating and simplifying the process, the work-flow analysis targets inefficiencies in the work motion. The work-flow analysis aims for the identification and elimination of unnecessary steps and the reduction of burdensome activities.

Work-flow analysis steps

Work-flow analysis involves the following steps:

1. Define the process in terms of purposes, objectives, and start and end points.

2. Identify functions of the organization or facility.

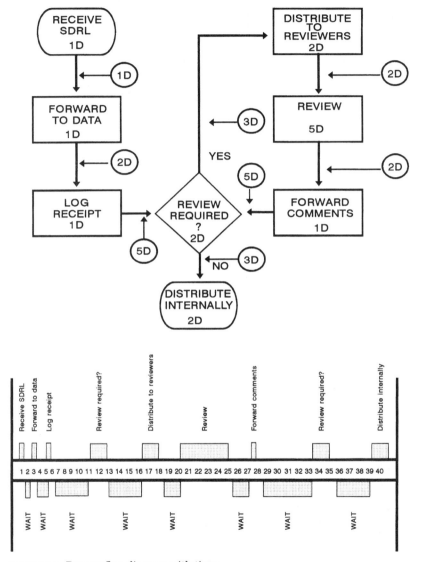

Figure 8.25 Process flow diagram with times.

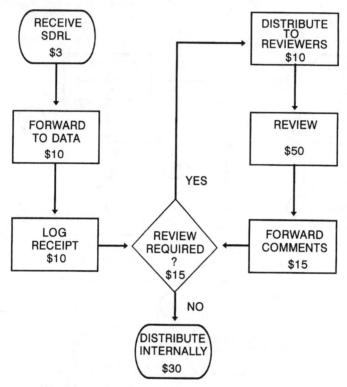

Figure 8.26   Process flow diagram with costs.

3. Identify activities within each function.

4. Identify tasks or basic steps within each activity.

5. Using process diagram symbols or drawings of the organization or facility, graphically display the actual work flow.

6. Analyze the work flow by identifying major activities, lengthy or complex tasks, decision points, duplicate or repetitive tasks.

7. Check the logic of the work flow by following all possible routes through the organization and facility for all work activity to ensure that all possible alternatives are explored.

8. Determine improvement opportunities.

Work-flow analysis example

The work flow for processing a customer's request for training within a learning center is shown in Fig. 8.27. The customer request is received by the administrative specialist at the desk. The administrative

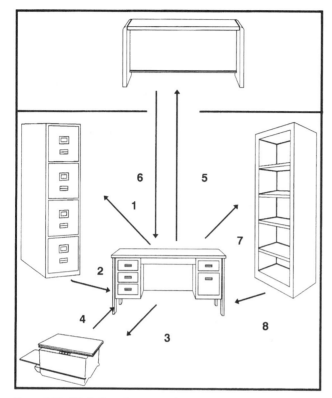

Figure 8.27 Work flow for processing a customer request and for training.

specialist goes to the file cabinet to get the person's training record. The training record is brought back to the desk to be checked to ensure that the customer's request meets the person's competency needs. Next, the administrative specialist makes a copy of the customer's request. The work-flow diagram shows the administrative specialist going to and from the copier. Once the administrative specialist determines that the customer's request requires enrollment in a training course, the course enrollment needs approval by the training manager. The administrative specialist takes the customer request with course enrollment information for signature. The administrative specialist returns to the desk. Finally, the customer's request is put in the course enrollment books on the bookcase. Because of this examination of the work flow, the customer request processing can be improved to save time and motion. For instance, duplicate copies could be eliminated, saving the time and motion to make the copies. The file cabinet and the bookcase could be moved closer to the desk to reduce motion.

Five Whys

The five whys is a powerful tool to find the "root" cause of a problem or situation quickly. It consists of repeating the question Why? until the underlying cause or causes are discovered. There is no magic involved with the five whys. In reality, discovering an answer could involve asking more or fewer Whys. This tool focuses on the process instead of the person by asking Why? instead of Who? In the majority of cases, the cause is the process. Only a small percentage of the time is the cause a person.

Five-whys steps

1. Describe the problem as specifically as possible.
2. Ask why the problem happens.
3. If the answer does not find the root cause, ask Why? again.
4. Repeat asking Why? until the root cause is identified.
5. Ask at least five Whys? before asking Who.

Five-whys example

A manager wants to find out why the customer's order was not shipped on time.

We missed the shipment time for the customer's order. WHY?

The order was not sent to order picking in time. WHY?

The sales order was not complete. WHY?

The customer's credit was being checked. WHY?

At this point, the manager recognizes the credit check as being the root cause of the problem. The manager decides to further analyze the credit check operation to speed up the process.

Cause-and-Effect Analysis

Cause-and-effect analysis is a useful technique for helping a group examine the underlying causes of a problem. Figure 8.28 shows a basic cause-and-effect diagram. The cause-and-effect diagram is a graphic representation of the relationships among a list of issues, problems, and opportunities. It is a useful tool in association with brainstorming because brainstorming is necessary to discover the core issues or root causes. Application of the analysis usually results in a more specific definition of the underlying cause of the problem. The technique has the added benefit of being very graphic, which helps members see patterns and relationships among potential causes. It lets them ex-

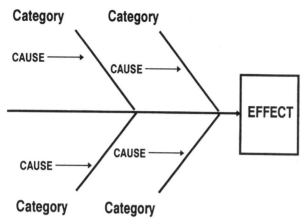

Figure 8.28   Basic cause-and-effect diagram.

press their interpretation of the nature of the problem; it also often stimulates further brainstorming and clarification of the problem, which leads to establishing priorities and taking corrective action.

For a process to be improved or a problem to be solved, the action taken must target the real issue, the underlying or root cause. It must attack the disease, not merely the symptoms. Cause-and-effect analysis aims at the cause of the issue or problem bringing the undesirable effect. The cause-and-effect diagram graphically displays how causes and the effect are associated.

The cause-and-effect diagram resembles a fishbone, as shown in Fig. 8.28. To make one up, start by drawing a box; inside, write the effect—the issue or problem. Next, draw an arrow pointing to the box. From the arrow, draw slanting lines, starting with four of them. At this point the team decides on categories. As the categories are selected, lines can be added or not used. There can be as many major categories as necessary. Typically, three to five categories are selected which may include manpower (people), materials, methods, machines, environment, culture, systems, procedures, training, plans, facility, policies, technology, information systems, communications, and structure. The categories are written above the slanted lines.

Once the categories are selected, the team brainstorms all the possible causes of the effect within each category. The brainstorming rules outlined in Chapter 6 apply for the cause-and-effect analysis. During the identification of causes, team members build on one another's ideas. The same cause can be in more than one category. For instance, the cause "unavailable" may be appropriate for several categories: people, machines, methods, and materials. The causes are written as little horizontal branches connected to the major-branch slanted line.

Once the causes are listed, the team interprets the diagram. The team looks for the underlying cause or causes by critically examining the pros and cons of each cause. In addition, the team should repeatedly ask why, which will provide a basis for focusing on consensus of selecting the underlying or root cause. This is accomplished by using selection techniques as described in Chapter 8.

Next the selected underlying cause or causes must be validated by collecting data on the occurrences of the causes. Again, the focus is a cure for the root disease, not just the symptoms. The cause-and-effect analysis steps can be summarized as follows:

1. Define the problem.

2. Define the major categories.

3. Brainstorm possible causes.

4. Identify the most likely causes.

5. Verify the most likely cause.

Cause-and-effect analysis example

Figure 8.29 shows the cause-and-effect diagram for the example. The steps of the cause-and-effect example are described in the following paragraphs.

Define the problem.   The team is asked to identify the problem, and the problem is the effect. In the example the problem is a piece of

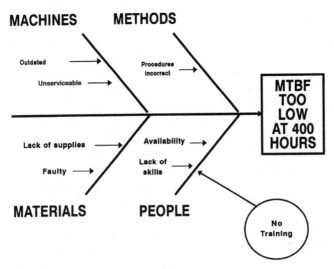

Figure 8.29   Cause-and-effect example.

equipment with a mean time between failure (MTBF) of 400 hours. This MTBF is too low. The desired goal is a MTBF of over 800 hours. Thus, the MTBF is the problem or effect.

Define the major categories. Next, the major categories of possible causes of the problem are identified. The most popular categories are the machines, methods, people, and materials. These categories were selected for the example. It is important to tailor the categories to the specific problem. Remember, you are not limited to these categories.

Brainstorm possible causes. The team then brainstorms possible causes. These causes are listed under the appropriate category. The brainstorming rules described in Chapter 6 apply in this step. It is sometimes helpful for the leader to keep repeating the heading for the cause in relation to the effect. For instance, under method, what is a cause of low MTBF? What is another method causing the low MTBF? This questioning is continued in each category until all ideas are exhausted. If someone comes up with an idea that applies to more than one category, list the idea in each category. If someone comes up with an idea that falls into a category other than the one being brainstormed, list the idea in the appropriate category. If an idea is generated that the person cannot immediately categorize, list the idea to the side of the diagram. The idea can be categorized at the completion of the brainstorming session. If someone cannot generate an idea, the person can pass or build on other people's ideas. Continue the brainstorming session until the team is satisfied they have completed their search for the underlying causes for the problem.

Identify the most likely causes. The team looks for clues to the most likely causes. Once all the causes are examined, the team selects by consensus the most likely cause by using the selection techniques found in this chapter.

Verify the most likely cause. The most likely cause is verified or rejected by data statistical analysis, a test, collection of more data on the problem, or communications with customers. When data statistical analysis is used, histograms or Pareto analysis are usually very effective in verifying the most likely cause. Once your team has identified the underlying cause, it can begin to generate alternative solutions and work toward improvement.

Data Statistical Analysis

Data statistical analysis is an essential element of any TQM endeavor. Customer-driven project management uses quantitative methods to continuously improve the project processes and all the processes in the organization aimed at total customer satisfaction. Involved in this endeavor are the monitoring, analyzing, correcting, and improving processes using rational decision-making based on facts. Use of statistics is one method of establishing factual data. Statistics are used for many purposes in a TQM environment, including problem solving, process measurement, and pass/fail decisions. Statistics are useful in step 1 of the TQM improvement methodology to define the quality issues; they help quantify total customer satisfaction. In the understanding and definition of the process, statistics provide factual data on process performance. In the analysis of the improvement opportunity, statistics aid in the verification of underlying or root causes, determining the vital few, evaluating variation, and establishing correlation. In the taking of action, statistics provide data for robust design and capable processes. In the checking of steps and monitoring of results, statistics furnish a yardstick for decision making.

In problem solving, statistics help in understanding the problem, determining actual performance, focusing on the vital few problems, and verifying the root or underlying causes. In process measurement, statistics assist in determining the performance and the causes of variation in a process. Once the process performance is known, process improvement goals can be determined. In addition, variations in the process can be analyzed to determine the cause—common or special. Next, the corrective can be aimed at doing nothing, either eliminating or reducing as much as possible the variation in the process or improving the process. The first option should always be considered when a chance variation in a process is discovered. This option is often overlooked, resulting in a correction of the chance variation only increasing the variation. Second, eliminate special causes. Third, reduce common causes. Finally, target attention on continuous process improvement. In pass/fail decisions, statistics can provide the criteria for the decision.

Data statistical analysis includes tools for collecting, sorting, charting, and analyzing data to make decisions. A chart can make the process easier to understand by arranging the data so that comparisons can be made to focus on the right problems. Sorting and resorting the data can help the team focus on the most important problems and causes. Targeting smaller and smaller samples or categories funnels the data to the underlying causes. Even small improvements on the right problems can yield significant benefits.

Data analysis steps

1. Collect data
2. Sort

3. Chart data

4. Analyze data

Data collection

Data collection is the first step in data statistical analysis. It starts with determination of what data is needed. Sometimes the data required is already available. In these cases, all that is required is to sort, chart, and analyze the data. However, in many cases, the specific data required for data statistical analysis is not available. If this is the case, the team needs to determine what data to collect, where to collect it, and how to collect it.

Data collection plan

Data collection requires a plan. The data collection plan establishes the purpose, strategy, and tactics to get the data for data statistical analysis. The data collection plan answers the following questions:

Why does the team need to collect the data?

What data is needed?

What process provides the data?

Where in the process is the data available?

Is the data already being collected?

If the data is not already being collected, how will it be collected?

Who will collect the data?

How long will the data be collected?

What data collection method will be used?

What sampling method is needed?

Who will chart the data?

How will the data be reported/presented?

Is a pilot or test necessary?

How will the pilot or test be conducted?

Who will participate in the pilot or test?

Is the data timely, accurate, and consistent?

Data collection methods

Data must be collected to measure and analyze a process. There are many methods for data collection. The data collection method must accomplish the purpose as stated by the customer-driven team in the data collection plan. Data collection methods include the following:

- Observation
- Questionnaire
- Interview
- Tests
- Work samples
- Checksheets

Observation

Observation is looking at actual performance or data, which could include reviewing documentation. This type of data collection is useful when one wants to distinguish between perceived and actual outcomes or behaviors.

Questionnaire or survey

A questionnaire requests in writing some particular information. A questionnaire is most useful when information is needed from a large number of people in a short time or when the people with the information are geographically distant. Questionnaires have some disadvantages, as respondents may misinterpret the questions, the returns could be low, and there is no opportunity to probe deeper into responses.

Interview

An interview involves personally communicating directly to the people with the information. Interviewing is used to collect information needed to improve processes and solve problems and to involve those outside the group in generating and implementing potential solutions. Interviewing is also useful when implemented solutions are being evaluated. Interviewing is essential for supplier and customer analysis. The disadvantage to an interview is that many times the outcomes depend on the skill of the interviewer. In addition, respondents may be hesitant to discuss personal, sensitive, or confidential information. In these cases, provisions must be made for guaranteed anonymity and confidentiality.

Tests

Tests measure outcomes. A test is useful when you need to measure specific results. Tests can be of people or items. A test can measure a person's knowledge on a subject. A test can also be used to determine if an assembly operates properly.

Work samples

Work samples involve checking or inspecting specific work outputs or work-in-process. Work samples are most useful in analysis of actual work performance.

Checksheets

Checksheets verify accomplishment of procedures. Checksheets are useful for qualifying or certifying performance exactly as specified.

Data collection charts

A data collection chart provides a means to document the information. A data collection chart simplifies data collection. An example of a simple, easy-to-use data collection chart is shown in Fig. 8.30. This data collection chart shows the discrepancies encountered in a part over a three-week period. As a discrepancy is encountered, the faulty item is checked. For instance, the board had 3 discrepancies in week 2, 1 in week 1, and 2 in week 3, for a total of 6 discrepancies for the three-week period. Look at the bottom of the chart to the component discrepancies in the first week; 15 discrepancies in the part were attributed to a faulty

DISCREPANCY	WEEK 1	WEEK 2	WEEK 3	WEEK 4	TOTAL
Board	/ / /	/	/ /		6
Solder	/	/			2
Wiring	/ / / / / /	/ / /	/ / / /		12
Connector	/ / / / /	/ / / / /	/ / / / /		15
Component	/ / / / / / / / / / / / / / /	/ / / / / / / / / / / / / / /	/ / / / / / / / / / / / / / /		45
TOTAL	31	25	24		80

Figure 8.30 Data collection chart.

component. Again in the second week, 15 discrepancies in the part were attributed to a faulty component; and in the third week, 15 more, for a total of 45. The totals for all discrepancies for the three weeks are

Board	6
Solder	2
Wiring	12
Connector	15
Component	45
Total all weeks	80

Data collection charts help to systematically collect data. The validity of the data depends on the collection of the data. The customer-driven team must ensure that the data is unbiased, accurate, properly recorded, and representative of typical conditions.

Data collection sampling

When data are collected for analysis, it is often impractical to check 100 percent of the items—the entire population of the data. A sample of the whole population may be all that is required. By taking a sample, reliable information can still be collected. A sampling table can help you determine an appropriate sample size. These tables are usually available from the industrial engineering, quality, or management services personnel in an organization. To reduce the chance for biased results, use a random or systematic method to select samples. A random sample allows each item an equal chance of being selected. A systematic sample selects every fifth, tenth, or twentieth item. This method of sampling reduces the chance for biased results.

Types of sampling

There are two common types of samples, nonrandom and random. Nonrandom sampling is accomplished using judgment. This type of sample cannot usually be verified. In a TQM environment, nonrandom sampling is only recommended as a predecessor to random sampling. Random samples are samples in which each item in the population has a chance of being selected. This type of sampling is most useful in a TQM environment. Random sampling assures accurate statistics. Two of the most common forms of random sampling are simple random sampling and stratified sampling.

Simple random sampling

Simple random sampling can be accomplished by using a list of random digits or slips. The random-digit method uses a number to represent the items in the population. For instance, imagine that the population consists of 80 items. The items are numbered 1 to 80. Next, the sample must be selected. The sample is selected by the use of a random-number generator or a table of random digits. Suppose the sample consisted of eight items. The random-number generator or table of random digits would select eight items indiscriminately. With the slip method, each item in the population is also numbered. The numbers are recorded on slips, and the slips are put in a box. The sample is drawn from the slips in the box. These random-sampling tools ensure that the sample represents the population so inferences can later be made about the population from sample data.

Stratified sampling

Stratified sampling divides the population into similar groups or strata. There are two methods of stratified sampling. In one method, a certain number of items are selected at random from each group/strata according to the proportion of that group/strata to the population. In another method, a certain number of items are selected from each group/strata, then the group/strata is given weight according to the proportion of the group/strata to the population.

Central limit theorem

The central limit theorem is stated thus: The mean (average) of the sampling distribution of the mean will equal the population mean (average) regardless of sample size; as the sample size increases, the sampling distribution of the mean will approach normal, regardless of the shape of the population. The central limit theorem allows the use of sample statistics to make judgments about the population of the statistic.

The central limit theorem is an important concept in statistics since checking the entire population is often impractical or impossible.

Types of data

The two types of data are variable data and attribute data. Variable data is data that can be measured, which involves characteristics having a range of values—quality characteristics such as thickness, width, temperature, force, wear, strength, and sensitivity. Variable data is characterized as nominal best, smaller best, and larger best. Attribute data is data that can be counted or classified. Attribute data is associated with such characteristics as pass/fail, have/have not, go/no go, grade (A/B/C), and accept/reject.

Data arrangement

Once the data is collected from the population or sample, it needs to be arranged in a meaningful way. The arrangement allows observation of such things as the highest and lowest values (range), trends, central tendencies, patterns, the values appearing most often, special causes, common causes, and so on. The arrangement of data helps to determine the measures of central tendency and the frequency distribution.

Measures of central tendencies

The measures of central tendencies are mean, median, and mode. The mean is the average of something. The median is the middle value. The mode is the value most often represented in the data. In a normal distribution, the mean, median, and mode are equal. Figure 8.31 shows a normal distribution bell-shaped curve.

Measures of central tendencies examples

The sample data is as shown. The left column is the number of the sample. The center column is made up of the ages of the people in the sample as collected. The right column is sample data arranged from lowest to highest.

1.	30	30
2.	33	33
3.	42	35
4.	38	36
5.	50	38
6.	42	42
7.	35	42
8.	47	47
9.	36	48
10.	48	50

Mean
Median
Mode

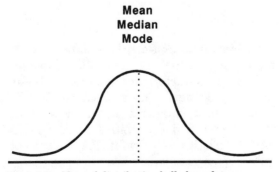

Figure 8.31   Normal distribution bell-shaped curve.

The average is calculated by summing all the items and dividing the sum by the total number of items, as follows:

$$\text{Mean} = \frac{30 + 33 + 35 + 36 + 38 + 42 + 42 + 47 + 48 + 50}{10}$$

$$= \frac{401}{10}$$

$$= 40.1$$

The median is the central item in a set of data. Half of the items fall above and half of the items fall below the median. In the example, the median is 40.

Mode is the value that is most often repeated in a set of data. In the example, the mode is 42.

Data charting

Once sorted, data must be put on a chart. Charts are pictures of the data that highlight important trends and significant relationships. Charts present the data in a form that can be quickly and easily understood. Charts serve as a powerful communications tool and they should be employed liberally to describe performance, support analysis, gain approval, and support and document the improvement process.

When using charts and graphs, label titles and categories for clarity. Keep it simple, and report all the facts needed to be fair and accurate.

There are many different types of charts or graphs available and useful in the TQM process. Some of the most common are the bar chart, pie chart, simple line chart (time plots or trend chart), histogram, and scatter chart. In addition, Pareto charts, control charts, and process capability charts are helpful for many specific TQM activities.

Bar chart. A bar chart is useful when comparisons are made between and among many events or items. Figure 8.32 shows a bar chart of the information from the checksheet. The bar chart shows the number of repairs by categories arranged from the highest to the lowest number of repairs.

Pie chart. A pie chart shows the relationship between items and the whole. Figure 8.33 shows a pie chart of the information from the checksheet. Displayed is each repair item's contribution to the total number of repairs. This chart clearly shows that component repair is the most common type of repair. In this case, the organization should focus efforts to reduce the number of failures in the components.

Line chart. A line chart is used when quantifiable information is de-

NUMBER

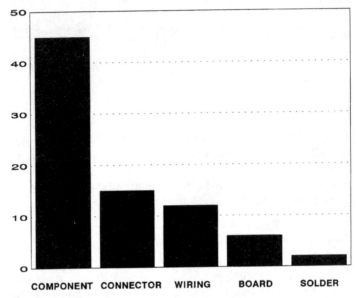

Figure 8.32 Bar chart.

scribed and compared. A line chart provides insight into statistical trends, particularly over a specified period of time. Figure 8.34 shows, in time-chart form, a line chart of the information from the checksheet. This chart displays the number of repairs by category per week. It shows the discrepancies among the number of repairs over the three-week period.

Scatter chart. A scatter chart and its related correlation analysis permit two factors and the relationship that exists between them to be compared and contrasted at the same time. A graphic display can help reveal possible relationships and causes of a problem even when links between the two factors are not evident. The pattern or distribution of the data points in a scatter diagram indicates the strength of the relationship between the factors being examined. It also indicates the type of relationship, whether it is positive, negative, a curve, or no relationship at all.

Figure 8.35 shows a series of scatter charts. These charts are displaying the relationship between the number of items tested and number of failures. The X axis of the chart is the number tested, and the Y axis is the number failed. First, the information is plotted on the graph. Next, a line is fitted through the scatter diagram. Then the

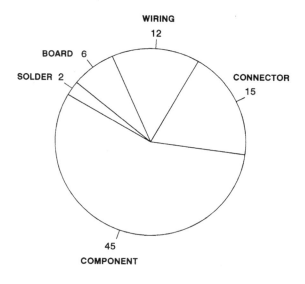

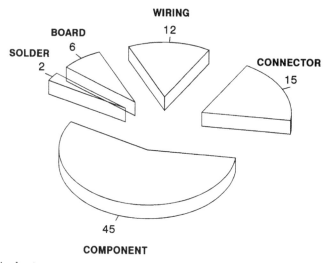

Figure 8.33   Pie chart.

chart is analyzed to determine the relationship. A positive correlation indicates a direct relationship between the factors, i.e., as the number of items tested increases, the number of failures also increases. A negative correlation shows an inverse relationship, i.e., as the number of items tested increases, the number of failures decreases. The relationship could be depicted by a curve, which is used when the number of

NUMBER

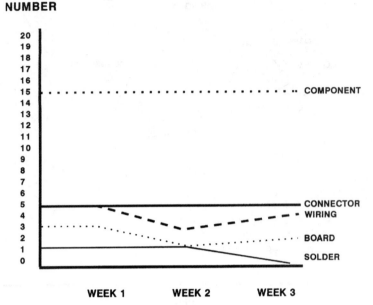

Figure 8.34   Line chart.

repairs changes by some fixed proportion. This type of curve is some-times referred to as a "learning curve." If no pattern is evident, no cor-relation exists.

Although a scatter diagram indicates a relationship between two factors, additional correlation analysis is usually required to substan-tiate the nature of the indicated relationship.

Histogram.   A histogram is a vertical bar chart that shows frequency of data in column form. The columns may be presented vertically or horizontally. Figure 8.36 shows a histogram. This type of data chart-ing is useful in identifying changes in a process. A histogram can pro-vide insight into the performance of a process and appropriate correc-tive actions by examining its centering, width, and shape. The closer the columns of the histogram are to the center of the chart, the more the process is on target. The wider the spread of the columns from the center, the greater the variation of the process from the target. Any change from a normal bell shape may indicate a problem area. Figure 8.37 shows some examples of various distributions.

Histogram example frequency distribution.   A histogram is usually a chart of frequency distribution. Frequency distribution is a table showing the number of elements in each class of a set of data. It

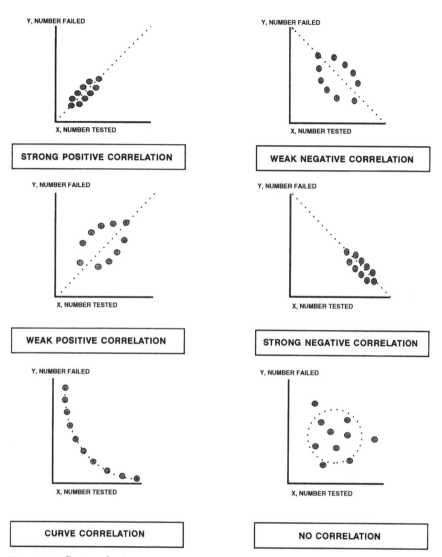

Figure 8.35   Scatter chart.

arranges data into classes with the number of observations in each class. Classes are groups of data describing one characteristic of the data. A frequency distribution displays the number of times an observation of the characteristic falls into each class.

In constructing a histogram, the first step is to collect the data. The raw data for the average inventory of work-in-process of one assembly area over a 15-day period is as follows:

Frequency

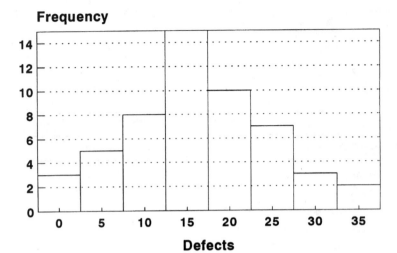

Figure 8.36 Histogram.

1	1	5	2	4	2	3	1	2	2
3	4	3	2	2					

The second step is to arrange the data from lowest to highest:

1	1	1	2	2	2	3	3	3	3
3	3	4	4	5					

The third step is to determine class intervals. The class intervals should be equal. One method for determining class intervals is highest value minus lowest value divided by number of classes. For the example, the formula provides a class interval of .8 being used. This gives the following class levels:

1	–	1.8
1.8	–	2.7
2.8	–	3.6
3.7	–	4.5
4.6	–	5.3

The fourth step is to sort the data into classes and count the number of points, as follows:

1	–	1.8	3
1.8	–	2.7	3
2.8	–	3.6	6
3.7	–	4.5	2
4.6	–	5.3	1

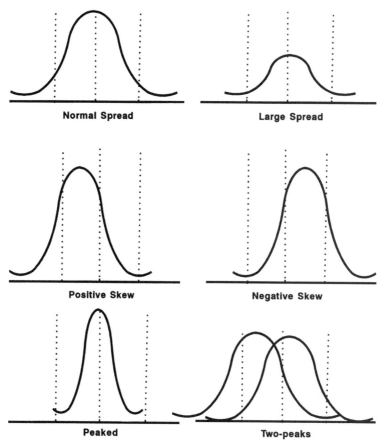

Figure 8.37   Example of various distributions.

The fifth step is to determine relative frequency and/or cumulative frequency of the data. The following array shows relative frequency and cumulative frequency of the example data:

				Relative		Cumulative
1	–	1.8	3	20	20	
1.8	–	2.7	3	20	40	
2.8	–	3.6	6	40	80	
3.7	–	4.5	2	13	83	
4.6	–	5.3	1	7	80	
			15	80	80	

The sixth step is to display the data on a histogram chart.

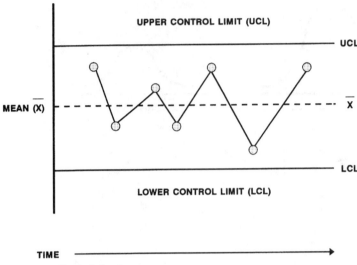

Figure 8.38 Control chart.

Control chart. A control chart displays the process performance in relation to control limits. Figure 8.38 shows a control chart. The control chart displays the data over time and shows the variation in the data. Control charts illustrate this variation.

Control charts are used to show the variation on a variety of variables. The control chart allows you to distinguish between measurements that are within the variability of the process and measurements that are outside the normal range and are produced by special causes. For instance, it may be that University College accepts that 80 percent of student complaints are being resolved after the first call. If data suggests that only 70 percent are being resolved after the first call, this number is outside the normal range.

The upper and lower parts of the normal range are referred to as the upper and lower control limits. These upper and lower control limits must not be confused with specification limits relating to acceptability of the process output. Control limits describe the natural variation of the process. Points within the limits are generally indicative of normal and expected variation.

Points outside the limits signal that special attention is required because they are beyond the built-in systemic causes of variation in the process. It is necessary to investigate only those points outside the control limits. The variation in the data is due to root causes, either common causes or special causes. Common causes are always present and the resulting variation in the process output will remain stable within the control limits. Special causes, or systemic problems, will result in data falling outside the control limits. This type of chart helps you to understand the inherent capability of your processes, bring your processes under control by eliminating the special causes of variation, reduce tampering with processes that are under statistical control, and monitor the effects of process changes aimed at improvement. Control charts are the fundamental visual display for statistical process control.

Statistical process control is used when a product or service is consistent, meaning that the process is stable or the process outcomes usually fall within prescribed specifications. Statistical process control and control charts are useful when some characteristic of a process output has changed from previously known and predictable levels or when the process has changed and the new upper and lower control limits need to be established.

There are many types of control charts. Some of the many types of control charts for variable and attribute data are listed below.

Variable charts:

Name of chart	What it measures
X-bar	Central tendency
R	Range
Sigma	Standard deviation

Attribute charts:

Name of chart	What it measures
p	Proportion nonconforming of total
c	Defects per subgroup for constant sample size
u	Percentage of nonconformities for varying sample size
np	Number nonconforming compared to total number

Analyzing the Data

Once the data has been collected, sorted, and put on charts, the data is analyzed to identify significant findings.

- Ask specific problem identification questions with "what," "when," "where," "who," "how much," "what are the causes," and "what is the impact?"
- Identify underlying causes.
- Clarify expected outcome.

Pareto analysis

One specific type of analysis is a Pareto chart. In the late 1800s, Vilfredo Pareto, an Italian economist, found that typically 80 percent of the wealth of a region was concentrated in less than 20 percent of the population. In recent times, Joseph Juran formulated what he called the Pareto principle of problems: only a vital few elements (20 percent) account for the majority of problems. The Pareto principle states that a large percentage of the results are caused by a small percentage of the causes. This principle is sometimes referred to as the 80/20 rule.

Some examples are as follows:

"20 percent of university students account for 80 percent of complaints."

"15 percent of errors produce 85 percent of the scrap."

The exact percentage is not important. In many cases, the ratios may be 8/80, 15/85, 30/70, or even 40/60. The Pareto principle has been proven in numerous situations with various results. The importance of this rule is to focus on the vital few problems that produce the big results instead of the trivial many that provide minor results. Greater success is probable if you do so.

Figure 8.39 shows a Pareto chart. The Pareto chart is simply a bar chart with the data arranged in descending order of importance, generally by magnitude of frequency, cost, time, or a similar parameter. The chart presents the information being examined in the order of priority and focuses attention on the most critical issues. The chart aids the decision-making process because it puts issues into an easily understood framework in which relationships and relative contributions are clearly evident.

The Pareto chart is a simple diagram of vertical bars showing first things first. The biggest or most important items are shown on the left with other items arranged in descending order to the right. The horizontal line indicates what and the vertical shows how much.

Pareto charts focus the team on the vital few areas during analysis. They can also be used to measure progress after a solution has been implemented. In addition, they can be used to show results on two separate charts showing before and after data. Also, Pareto charts are useful to display consensus during problem solving.

Figure 8.39 shows the construction of a Pareto chart. First, data is collected using a data collection worksheet. In the example, data was collected on five potential underlying causes of engineering changes. The frequency data is listed in the first column of the Pareto chart worksheet. Second, the percentage of the total for each cause is calcu-

Pareto Chart Worksheet

	Causes	Frequency (or other measure)	Percentage of total	Cumulative percentage
Cause 1	Performance	30	46	46
Cause 2	Cost	15	23	69
Cause 3	Producibility	10	15	84
Cause 4	Quality	5	8	92
Cause 5	Supportability	5	8	100

PARETO CHART

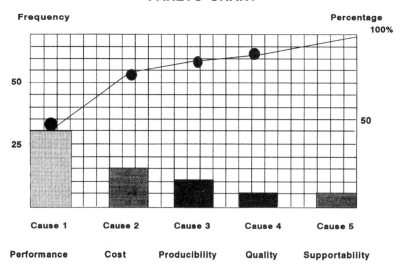

Figure 8.39   Pareto chart.

lated. The results of this calculation are annotated in the second column of the Pareto chart worksheet. Third, the cumulative percentage is computed. This is shown in the third column of the Pareto chart worksheet. Fourth, the chart is drawn on graph paper, with the horizontal axis being the causes and the vertical axis the frequency. Fifth, the graph is scaled. Put zero at the bottom and the total at the top. Mark equal intervals in between the bottom and top. Sixth, arrange the causes from highest to lowest. Seventh, construct the bar chart. Start with the cause with the highest number on the left and work right in descending order to the lowest cause. The height of each bar indicates the number of times (frequency) that cause was counted. Eighth, indicate the percentages on the right of the chart. Ninth, put dots to mark cumulative percentages from the Pareto chart worksheet. Tenth, analyze the chart. In this chart, the vital few are performance, cost, and producibility. By focusing on correcting these problems, you can eliminate over 80 percent of the engineering changes.

Variability analysis

As mentioned before, variability exists in everything. The presence of variability is a major obstacle to quality. Variation is quality's major enemy. For example, if you weighed yourself 10 times, there would be some variation in the measurements. The actual weight is probably near the average of all measurements. Variation has common and special causes. By examining the statistical data using statistical process control, you can monitor, control, and improve deviations from target values. Variability analysis is an essential tool of TQM.

Process Capability Analysis

Process capability analysis provides an indication of the performance of a process. It involves measuring process performance in relation to being able to produce the process output within engineering specifications. This analysis is accomplished through use of process capability indexes. Two of the most common process capability indexes are the CP and CPK indexes. The CP index gives the ratio of the specification limits to the process limits. This figure shows whether the process is capable of producing within specifications. The objective is a CP greater than or equal to 1. CPK is the location of the process range within the specification limits. This index provides evidence that the product meets specification. The objective is a CPK greater than or equal to 1. The process is centered when the CPK and CP equal 1, which indicates a capable process.

Force-Field Analysis

Force-field analysis is a technique that helps a group describe the forces at work in a given situation. The force-field analysis chart is shown in Fig. 8.40. The underlying assumption is that every situation results from a balance of forces: restraining forces and driving forces. Restraining forces are those elements that keep the situation from improving; driving forces are those elements that are pushing toward the achievement of the goal. Force-field analysis forces the

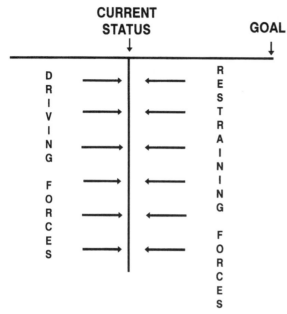

CURRENT
STATUS

GOAL

D
R
I
V
I
N
G

F
O
R
C
E
S

R
E
S
T
R
A
I
N
I
N
G

F
O
R
C
E
S

Figure 8.40   Force field analysis chart.

team to examine strengths as well as problems. Sometimes by building on a driving force or strength, a team can bring about the needed improvement.

Force-field analysis steps

1. Define the current status and goal.
2. Identify and prioritize the restraining forces.
3. Identify the driving forces for each restraining force.
4. Identify owners and the level of management best suited to correct the problem.
5. Use continuous improvement cycle.

Force-field analysis example

In the example shown in Fig. 8.41, the goal is to provide logistics training for engineers. The current status is that no training exists. The restraining forces are management support, funding, courseware, trainers, training equipment, and organizational culture. It is sug-

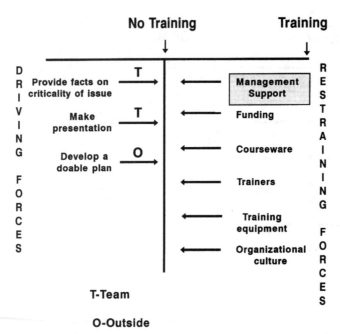

Figure 8.41   Force-field analysis example.

gested when performing a force-field analysis that the team select the predominant restraining force first. Then, if needed, other restraining forces can be analyzed, one restraining force at a time. The restraining force of management is selected to develop the driving forces. The team brainstorms the driving forces. In the example, the team determined that the driving forces that can be used to eliminate or weaken management's nonsupport are facts, communication, and a plan. Next, the team decides which alternatives the team can do. This is shown by T for team or O for outside the team's control. The team determines it can provide facts and make a presentation. The other alternative to develop a doable plan requires outside assistance. Finally, the team selects an alternative or alternatives to act upon.

Implementation Tools and Techniques

1. Establish what needs to be accomplished.
2. Determine who is responsible for performing the work.
3. Develop a schedule to complete the project.

Task Title	Duration	Start	End	Resources

Figure 8.42   Task list format.

4. Define resources required to complete the project, including people, funds, equipment, and supplies.
5. Manage the project's progress.

Task list

The task list includes the development of the tasks involved in the project. The task list forms the basis for project scheduling.

Figure 8.42 shows the basic task list format. It provides a documentation of

- *Task title.* The task. In some cases, it is just a name. In other cases, it may include a name and identification code.

- *Task duration.* The total amount of time necessary to complete the task.

- *Task resource.* The human resources, materials, and equipment to complete the task.

Task list construction

The task list can be formulated by forward or reverse planning. With forward planning, the customer-driven project lead team sets the start date and then lists the first activity to the last activity. With reverse planning, the team starts with the end activity and works backward to the first activity. This is sometimes called "backward mapping." It is usually advantageous to use reverse planning. Reverse planning focuses on the project's end result by always keeping the project's output objective in view. In reverse planning, the task list is constructed by first asking "What activities are required to complete the project's object?" and then repeatedly asking "What activities does it take to complete the preceding activity to complete the project's objective?" until all the activities to complete the project are identified.

Task list example

Figure 8.43 shows a typical task list for applying the process improvement methodology. It shows all the steps from the beginning to project close-out. Next, the task durations are determined. Then, the estimated start and end dates would be formulated. Finally, the resources to perform each task would be listed.

Project schedule

The project schedule involves the scheduling, monitoring, and managing of the project. Through use of the information from the task list, the project schedule organizes the project by describing the sequential

Task Name	Duration	Start	End
Implement improved process	135.00 d	Oct/30/95	May/10/96
Determine "as is" baseline	20.00 d	Oct/30/95	Nov/27/95
Develop plan	30.00 d	Nov/28/95	Jan/10/96
Present plan for approval	10.00 d	Jan/11/96	Jan/25/96
Improved process approval	0.00 d	Jan/25/96	Jan/25/96
Prepare preliminary procedure	5.00 d	Jan/26/96	Feb/01/96
Coordinate preliminary procedure	5.00 d	Feb/02/96	Feb/08/96
Validate new process	10.00 d	Feb/09/96	Feb/23/96
Finalize new procedure	15.00 d	Feb/26/96	Mar/15/96
Prepare training for new process	10.00 d	Mar/18/96	Mar/29/96
Validate training for new process	5.00 d	Mar/18/96	Mar/22/96
Publish new process procedure	5.00 d	Apr/01/96	Apr/05/96
Conduct training	5.00 d	Apr/08/96	Apr/12/96
Measure new process performance	15.00 d	Apr/15/96	May/03/96
Evaluate new process	5.00 d	May/06/96	May/10/96

Printed: Oct/29/95
Page 1

Figure 8.43 Typical task list for applying process improvement.

relationship of project activities and milestones. To do so, the end point of the project must be determined, the tasks that need to be accomplished must be defined, and the time when the project needs to be completed must be specified.

Besides the relationships between tasks, teams give their estimates of how long it will take to deliver the work after all events necessary to start it have been completed. A projection is made after schedules are compared with elapsed-time estimates. A comparison is also made of the estimates to the actual required date of completion. In the planning stage, teams can do any of the following actions:

- Revise activity duration times
- Revise relationships of tasks
- Revise technical objective

Once the time estimates meet the required project completion date, the next step is to determine resource requirements. The teams assign resources, including people, equipment, and materials, for each activity. Again, the project is evaluated to determine if the project can be completed on time. In this case, the teams can do any of the same actions used for time management as well as revise the resource allocation. In addition, the team must consider the cost and time constraints of the project.

Once the time and resource estimates meet the project's time, cost, and technical constraints, the project schedule baseline for monitoring the project's progress is established. During its performance, the project is continually evaluated based on the aforementioned triple constraints. Any variance in time, cost, or technical performance must be analyzed for possible corrective action.

Once under way, the project must be continually guided and managed through use of the critical path. The critical path represents the longest route between the first event and the last event. It sometimes changes over the course of the project, in which case the team can decide to (1) reestimate activity duration on the critical path; (2) revise network relationships by changing dependencies; (3) revise cost and/or resource objectives; or (4) revise the technical objective by changing the deliverable specification.

The project schedule should be seen as support for project planning, scheduling, monitoring, and management. It is a means to an end. It should never be allowed to drive the project so far as to become a dominating factor that replaces good judgment and perspective. There is nothing magic about the critical path itself, but it does provide a valuable frame of reference for monitoring the project and for identifying targets for close attention and possible decision by the team.

The specific use of project network scheduling is a function of the complexity of the project and the sophistication of the organization employing it.

The common bar chart, or Gantt chart, will show which activities precede which other activities, and it will indicate the importance of those activities that will take the longest time. For a simple project this information may be sufficient to plan, schedule, monitor, and manage the project. For most projects, a network schedule will provide the information required for successful project management.

Simple Gantt project scheduling

For most simple projects, including many TQM projects, a simple Gantt project schedule is sufficient for project planning, scheduling, and management. For example, suppose a customer-driven quality improvement team requires the implementation of an improved process. The implementation of the improved process is the project. The Gantt chart to implement the improved process is shown in Fig. 8.44.

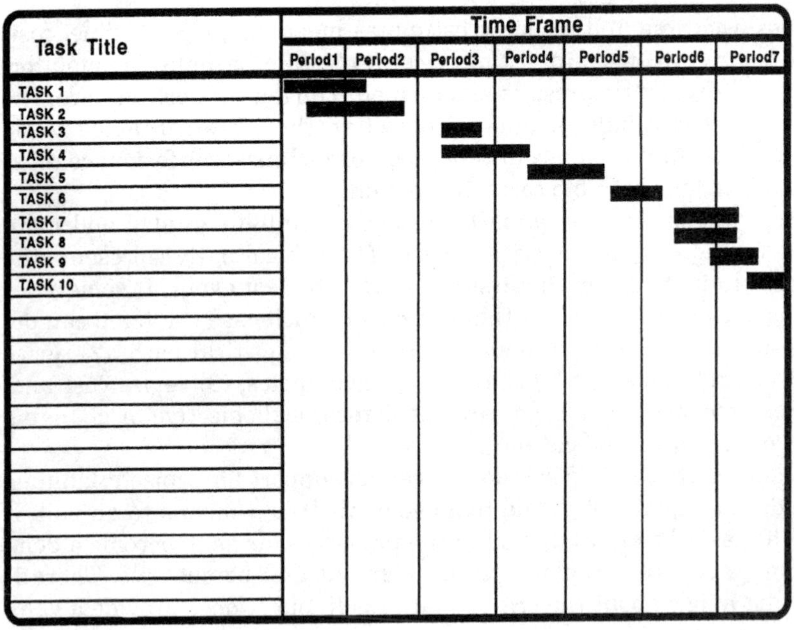

Task Title	Time Frame						
	Period1	Period2	Period3	Period4	Period5	Period6	Period7
TASK 1	▆						
TASK 2	▆▆						
TASK 3		▆					
TASK 4		▆▆					
TASK 5			▆▆				
TASK 6				▆▆			
TASK 7					▆▆		
TASK 8					▆▆		
TASK 9						▆▆	
TASK 10							

Figure 8.44 Gantt chart.

Chapter

9

Improving Systems: Advanced Tools and Techniques

Improved systems bring to market the best possible deliverables for achieving customer satisfaction at the optimum cost.

This chapter contains the following tools and techniques:

- Concurrent engineering
- Quality function deployment
- Robust design
- Statistical process control
- Cost of poor quality
- Miscellaneous other methodologies

System Improvement

Many times a system or process must be developed or completely redesigned to make an improvement. System improvement focuses on the development or redesign of systems. The system involved can be as complicated as an entire F-16 plane or a car or as simple as a flight-control surface or a car door. The system improvement tools and techniques can be used for any system, subsystem, or part. In fact, some of the tools, like statistical process control (SPC) and quality function deployment (QFD), have been used successfully for the continuous improvement of entire organizational systems.

Since the performance of a product is critical to customer satisfaction, this chapter focuses on the system improvement of a product.

The product is any output to a customer, including a system, subsystem, or part. A major impact on product performance is the product design, process design, and production processes.

The tools and techniques described in this chapter have specific application in the product design, process design, and production processes. However, these tools and techniques can be used to improve any system in the organization. They are applicable for improving whole systems, subsystems, or parts.

As stated earlier, system development/improvement focuses on improving the actual performance of a product through product design, process design, and planning of the production processes.

As shown in Fig. 9.1, system improvement starts with the customer. Next, the product and processes are designed. The voice of the customer carries through the product design and process design to the actual production of the product. Within the product design, process design, and production processes, specific tools are useful for ensuring customer satisfaction. The specific tools are concurrent engineering (CE), robust design (RD), quality functional deployment (QFD), statistical process control (SPC), and cost of poor quality (COPQ).

Concurrent engineering is useful during the product and process planning and design phases for reducing the time and cost of product

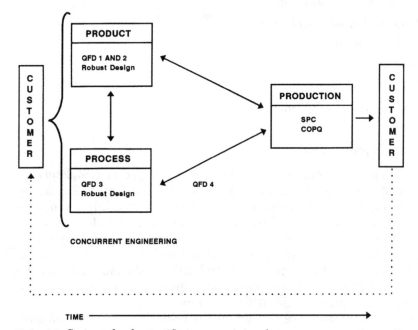

Figure 9.1   Systems development/improvement overview.

development. Quality functional deployment is beneficial for carrying the voice of the customer throughout the entire process. Robust design focuses on designing in quality by eliminating loss. Statistical process control is a technique for measuring process behavior during production. Cost of poor quality emphasizes eliminating waste in all the processes.

Concurrent engineering

Concurrent engineering (CE) is a philosophy and set of guiding principles in which product design and process design are developed concurrently—that is, with some product design and process development overlapping, including production and support planning. Figure 9.2 shows the difference between sequential engineering and concurrent engineering. With sequential engineering, the engineering phases are accomplished one after the other. Concurrent engineering overlaps the engineering phases.

Concurrent engineering is a subsystem of TQM which focuses on system and parametric design. (See the section in this chapter entitled "Robust design" for more detail.) Like TQM, concurrent engineering requires a management and cultural environment, teams, and an improvement system that focus on customer satisfaction.

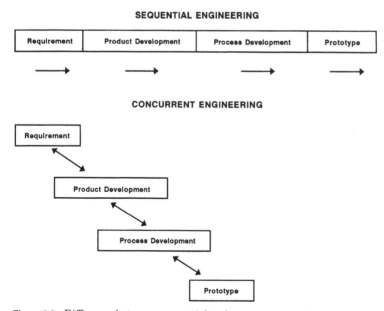

Figure 9.2 Difference between sequential and concurrent engineering.

The concurrent engineering philosophy emphasizes customer focus; advocates an organizationwide, systematic approach using a disciplined methodology; and stresses the never-ending improvements of product, processes, production, and support. It also involves the concurrent, simultaneous, or overlapping accomplishment of the phases of the project. For instance, the concept and design phases and the design and development phases are often accomplished simultaneously. Sometimes, also, activities in the development and production phases overlap. In most cases of concurrent engineering, all the phases contain some overlapping activities. Concurrent engineering requires upper management's active leadership and support to be successful. It accents robust design that decreases loss. It aims at reducing cost and time while improving quality and productivity. It uses the latest engineering planning initiatives including automation. Concurrent engineering forges a new reliance on multifunctional teams using tools and techniques like quality function deployment, design of experiments, the Taguchi approach, statistical process control, and so on.

A more formal definition from the Institute for Defense Analysis states: "Concurrent Engineering is a systematic approach to the integrated, concurrent design of products and their related processes, including manufacture and support. This approach is intended to cause the developers, from the outset, to consider all elements of the product life cycle from conception, through disposal, including quality, cost, schedule, and user requirements."

Concurrent engineering considerations

Customer focus

Organizationwide, systematic approach

Never-ending improvements in product, process, production, and support

Concurrent design of product and related processes

Upper management leadership

Robust design

Reduction in cost and time while quality and productivity are improved

Engineering planning initiatives, including automation

New reliance on multifunctional teams

Tools and techniques

Concurrent engineering steps

The concurrent engineering steps are

1. Establish a multifunctional team. Ensure representation from all required disciplines. The team should include representatives from such functions as systems/design engineering, reliability and maintainability engineering, test engineering, manufacturing engineering, production engineering, purchasing, manufacturing test and assembly, logistics engineering, supportability engineering, marketing, finance, and accounting.

2. Use a systematic disciplined approach. Select a specific approach using appropriate tools and techniques.

3. Determine customer requirements. Be sure to communicate with customers.

4. Develop product design, process design, and the planning of production and support processes together.

Quality function deployment (QFD)

Quality function deployment (QFD) is a disciplined approach for transforming customer requirements, the voice of the customer, into product development requirements. QFD is a tool for making plans visible and then determining the impact of the plans. QFD involves all activities of everyone at all stages from development through production with a customer focus.

Figure 9.3 shows the four phases of QFD. These phases are product planning, parts deployment, process planning, and production planning. The output from each phase is the input for the next phase. During phase 1, customer requirements are transformed into design requirements. In phase 2, design requirements are converted into a system (part) or concept design. Phase 3 examines candidate processes and selects one. Phase 4 looks at making capable production processes.

QFD House of Quality

The results of QFD planning are included on a chart called the "House of Quality." The basic House of Quality planning chart, which is for phase 1, is shown in Fig. 9.4. From this basic chart, many other useful charts are generated to assist in moving from general customer requirements to specific production processes in other steps of the customer-driven project management improvement methodology. The number and kinds of charts vary with the complexity of the project. QFD can be applied to the complete product, the system, the subsys-

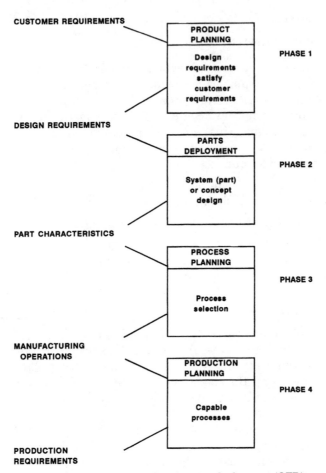

CUSTOMER REQUIREMENTS

PRODUCT
PLANNING

Design
requirements
satisfy
customer
requirements

PHASE 1

DESIGN REQUIREMENTS

PARTS
DEPLOYMENT

System (part)
or concept
design

PHASE 2

PART CHARACTERISTICS

PROCESS
PLANNING

Process
selection

PHASE 3

MANUFACTURING
OPERATIONS

PRODUCTION
PLANNING

Capable
processes

PHASE 4

PRODUCTION
REQUIREMENTS

Figure 9.3   Four phases of quality function deployment (QFD).

tem, and/or specific parts. At all stages of QFD application, prioritization is used to assure that the overall analysis does not become excessively burdensome in terms of time and cost.

QFD House of Quality steps for phase 1

Step 1: Determine the whats. These are the voice of the customer or customer needs and expectations.

Step 2: Transform the whats to hows. These hows become the product design requirements, or characteristics, which are measurable.

Step 3: Determine the nature of the relationships between the whats and the hows using a relationship matrix.

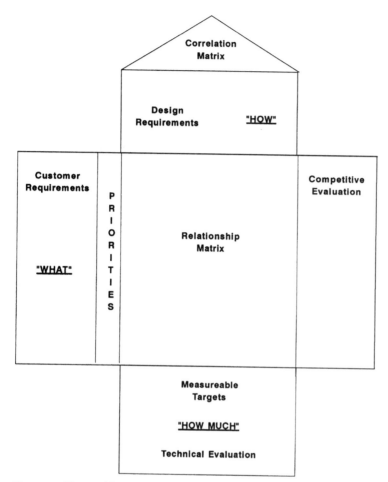

Figure 9.4 House of Quality planning chart.

Step 4: Establish the how much data, which provides target values for design requirements.

Step 5: Correlate each how to each other how, in the correlation section, or the roof of the houselike matrix. This step is used to aid in conflict resolution and trade-off analysis.

Step 6: Complete the two competitive evaluation sections. These competitive evaluations rate the product under question against similar products produced by the competition. One evaluation relates the product features to customer satisfaction, and the other evaluation assesses the product based on technical merits.

Step 7: Assign or calculate importance ratings to help prioritize analysis efforts.

Step 8: Analyze results. This step includes a check-and-balance procedure to both identify planning gaps and point to wasteful activities.

Quality function deployment example

Figure 9.5 shows an example of part of a QFD phase 1 chart for delivering an excellent cup of coffee. The customer requirements for an excellent cup of coffee are "hot," "eye-opener," "rich flavor," "good aroma," "low price," "generous amounts," and "stays hot." These are the "whats." The "hows" are the product requirements listed across the top of the matrix. These include: "serving temperature," "amount of caffeine," and so on. The relationship matrix consists of the "whats" along the left column and the "hows" across the top. The evaluation of the relationship between the customer requirements and the product requirements is shown by the symbols depicting a weak, medium, or strong relationship. Next, the "how much" is shown on the bottom of the matrix. The "how much" is determined by examining the "what" to the specific "how" to get the "how much". The "how much" items include: "120–140 degrees Fahrenheit," "_____ ppm," and so on. In the example, the "what" of "low price" is examined in relation to the "how" of "sale price" to get the "how much" of $0.40. The roof of the House of Quality is the correlation matrix. In the correlation matrix, trade-off analysis is performed by comparing each "how" to the others. These relationships can be strong positive, positive, negative, and strong negative, as indicated by the symbols.

The example also shows the evaluation sections for customer and technical competitive assessments. The customer competitive evaluation is along the right side of the House of Quality and the technical competitive evaluation is at the bottom portion.

In addition, the example displays the ratings for customer requirements, technical difficulty, and technical importance. The customer importance rating is next to the "what." The technical difficulty rating is above the "how much." The overall importance rating is shown on the bottom of the House of Quality.

Robust design

Robust design means design of a product having minimal quality losses. There are several methodologies associated with robust design. The major ones are traditional design of experiments (DOE) and the Taguchi approach. Traditional design of experiments is an experimen-

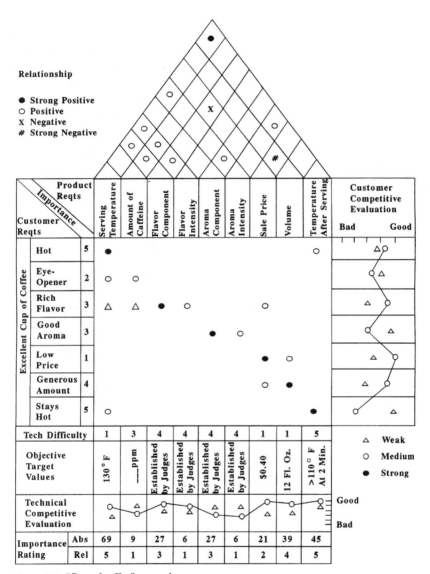

Figure 9.5 "Cup of coffee" example.

tal tool used to establish both parametric relationships and a product/ process model in the early (applied research) stages of the design process. However, traditional design of experiments can be very costly, particularly when it is desired to examine many parameters and their interactional effects. Traditional DOE examines various causes of performance for their contribution to variation, with a focus on arriving at the most influential causes of variation. Traditional design of experiments may be a useful tool in the preliminary design stage for modeling, parameter determination, research, and establishing a general understanding of product phenomena.

A major approach to robust design is the Taguchi approach. The Taguchi approach focuses on quality optimization. "Quality optimization" is based on Dr. Taguchi's definition of quality. Taguchi, in his book *Introduction to Quality Engineering,* states, "Quality is the (measure of degree of) loss a product causes after being shipped, other than any losses caused by its intrinsic functions." Simply put, any failure to satisfy the customer is a loss. Loss is determined by variation of performance from optimum target values. Loss, therefore, in the form of variability from best target values, is the enemy of quality. The goal is to minimize variation by designing a system (product, process, or part) having the best combination of factors, for example, centering on the optimum target values with minimal variability. By focusing on the bull's-eye, we make the product, process, or part insensitive to those normally uncontrollable "noise" factors that contribute to poor product performance and business failures. The Taguchi approach is not simply just another form of design of experiments. It is a major part of the successful TQM philosophy.

Loss function

The loss function is a key element of the Taguchi approach. The loss function examines the costs associated with any variation from the target value of a quality characteristic. As shown in Fig. 9.6, any variation from the target is a loss. At the target value, there is little or no loss contribution to cost. The farther from the target, however, the higher the costs. Costs get higher as values of the quality characteristic move from "best" to "better" to "poor" levels.

Robust design phases

In the Taguchi approach, the design of a product or a process is depicted as in Fig. 9.7. Product or process designs have three phases:

Systems (part) or concept design.   This phase arrives at the design

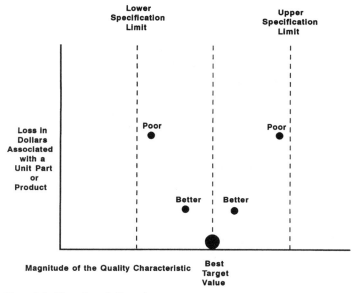

Figure 9.6 Target variation—loss.

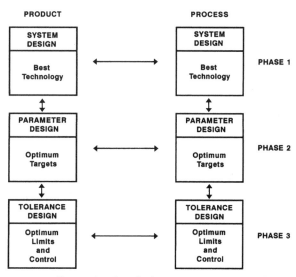

Figure 9.7 Process/product design.

architecture (size, shape, materials, number of parts, etc.) by looking at the best available technology.

Parameter (or robust) design. This stage focuses on making the
product performance (or process output) insensitive to variation by
moving toward the best target values of quality characteristics.

Tolerance design. This stage focuses on setting tight tolerances to
reduce variation in performance. Because it is the phase most responsible for adding costs, it is essential to reduce the need for setting tight tolerances by successfully producing robust products and
processes in the parameter design phase.

Cost of poor quality

Cost of quality is a system providing managers with cost details often
hidden from them. Cost of quality includes both the cost of conformance and the cost of nonconformance to quality requirements. Costs
of conformance consists of all the costs associated with maintaining
acceptable quality. The cost of nonconformance, or the cost of poor
quality, is the total cost incurred as a result of failure to achieve quality. Historically, organizations looked at all costs of quality. Today,
many excellent organizations are concentrating strictly on the nonconformance costs. This approach highlights the waste, or losses, due
to deviation from best target values. Once these costs are determined,
they can be reduced or eliminated through application of the continuous improvement philosophy.

Typically, the cost of nonconformance includes such items as inspection, warranty, litigation, scrap and rejects, rework, testing, retesting,
change orders, errors, lengthy cycle times, inventory, and customer
complaints.

Statistical process control

Statistical process control (SPC) is a statistical tool for monitoring
and controlling a process. SPC monitors the variation in a process
with the aim to produce the product at its best target values.

Figure 9.8 shows the major elements of statistical process control.
These elements are a process chart consisting of data plots, upper
control limit (UCL), lower control limit (LCL), and the mean for the
process.

Figure 9.8 also illustrates variation in a process. The variation is the
result of both common and special/assignable causes. Common causes
produce normal variation in an established process, whereas special/as-
signable causes are abnormal causes of variation in the process.

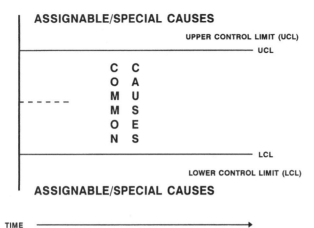

Figure 9.8   Statistical process control—major elements and variation.

Statistical process control steps

There are four steps in SPC:

1. Measure the process. Ensure that data collection is thorough, complete, and accurate.

2. Bring the process under statistical control. Eliminate special/assignable causes.

3. Monitor the process. Keep the process under statistical control.

4. Improve the process by seeking the best target value.

Miscellaneous Other Methodologies

There are many methodologies that can be used for system development/improvement. Some of the major methodologies have been discussed in the chapters in this book that concern themselves with tools and techniques. In addition, there are numerous other methodologies available to strive for success. Some of the more common systems include just-in-time (JIT); total production maintenance (TPM); mistake-proofing; manufacturing resource planning (MRP); computer-aided design, computer-aided engineering, and computer-aided manufacturing (CAD/CAE/CAM); computer integrated manufacturing (CIM); computer systems; information systems (IS); and total integrated logistics (TIL).

Just-in-time

Just-in-time (JIT) is a philosophy and methodology for having the right material just in time to be used in an operation. JIT reduces inventory and allows immediate correction of defects. This methodology is used for reducing waste, decreasing costs, and preventing errors.

Total production maintenance

Total production maintenance (TPM) is a system for involving the total organization in maintenance activities. TPM involves focusing specifically on equipment maintenance. TPM emphasizes involvement of everyone and everything, continuous improvement, training, optimum life-cycle cost, prevention of defects, and quality design. This methodology is effective for improving all production maintenance activities.

Mistake proofing

Mistake proofing, or poka-yoke, is a method for avoiding simple human error at work. The application of mistake-proofing frees workers from having to concentrate on simple tasks and allows them more time for process improvement activities. It is a major measure in the prevention of defects.

Manufacturing resource planning

Manufacturing resource planning (MRP) is an overall system for planning and controlling a manufacturing company's operations. MRP is used as a management tool to monitor and control manufacturing operations.

Computer-aided design, computer-aided engineering, and computer-aided manufacturing

Computer-aided design, computer-aided engineering, and computer-aided manufacturing (CAD/CAE/CAM) are automated systems for assisting in the design, engineering, and manufacturing processes. CAD/CAE/CAM are used to improve systems and processes, enhance product and process design, reduce the time factor, and eliminate losses.

Computer integrated manufacturing

Computer integrated manufacturing (CIM) is the integration of computer-aided design and computer-aided manufacturing (CAD/CAM) for all the design and manufacturing processes. The CIM methods improves on the CAD/CAM weapon system by eliminating redundancy.

Computer systems

Computer systems include a wide range of items such as hardware, software, firmware, robotics, expert systems, and artificial intelligence. Computer systems are a major technological methodology.

Information systems

Information systems (IS) are an automated system used to focus an organization toward its vision. An information system is used to plan, design, analyze, monitor, and respond to critical strategic information essential to achieving customer satisfaction (internal/external). An information system allows continuous review, analysis, and corrective action.

Total integrated logistics

Total integrated logistics (TIL) is the integration of all the logistics elements involved in the inputs to the organization, all the processes within the organization, and the outputs of the organization to ensure total customer supportability at an optimum life-cycle cost. This method aims at total customer satisfaction by supporting the operations of the organization and the customer. Total integrated logistics can be a major differentiator.

10

People Development: Tools and Techniques

Performance = Ability + Willingness

Organizational success comes to individuals working in a positive organizational environment, with effective and efficient systems, using appropriate technology. Ultimately, the performance of the people in the organization can make or break it. To be successful, Total Quality Management must be made relevant on an individual basis so that desired performance is achieved. In the VICTORY model, the "T" for training, "O" for ownership," and "R" for recognition and rewards all impact performance. The "T" relates to developing the ability of people. The "O" and "R" target the willingness part of the equation. Chapter 3 describes each of these elements.

Although this chapter will focus on individual development, it is important to keep in mind that the entire TQM process is an organizational development initiative. Therefore, an understanding of organizational development is essential for TQM to be successful. People development, too, is a key to success of the TQM process. Therefore, it is important to ensure integration of both the organizational development and people development interventions to achieve a common purpose. While people development goes from one individual at a time to improvement in the whole organization, organizational development targets improving the organizational systems. They must both be focusing on a common target.

Individual development requires there to be a disciplined process focused on the competence of the organization, with knowledge and skills instilled and application and experience granted by various methods. The organizational development process requires utilization of coaching, facilitating, and training interventions. In addition, it demands that there be management, leadership, support, and empowerment. In the beginning of the TQM effort, direction, communication, and focus are critical. Strong management and leadership are essential. In the next stage, people need help in understanding and doing what is expected; they get this assistance from coaching, training, and facilitating. Later, as they are using the TQM process, it is imperative that they be provided with additional support. Finally, there must be confidence for empowerment. Let the people closest to the process perform and improve their process.

This chapter focuses on the specific action process for people development in any organization, with the objective being the creation of a continuous learning organization. The ultimate advantage of any organization comes from being a learning organization, one that has the capability to adapt to an ever-changing environment. Attainment of such a goal requires there to be a high degree of both organizational and individual competence.

People Development

> Improvement begins with "I."

The ability of the people in an organization is a key element of the success of a TQM effort. Once TQM is implemented, the roles of everyone in the organization change. Executives take on a more visionary role than they did before. Managers become leaders. Workers add improvement responsibilities. This transformation requires new competencies to be instilled. The type of competencies needed will vary by organization. In addition to this variance, the competencies will be ever-changing, depending on the organizational environment in which they will be utilized. These facts mean that each organization needs to become a learning organization, and each individual needs to constantly grow and develop, to keep pace and to remain competitive.

Generally, today, key competencies embrace the integration of leadership, membership, teamwork, business systems, TQM, and technical systems. This relationship is shown in Fig. 10.1. Within each of these areas, each organization needs individuals to possess certain competencies so that success can be achieved.

Individual competencies

An individual can become "competent" when he or she is able to take certain knowledge and skills and successfully apply them in a given organization. Knowledge and skills are demonstrated by behaviors

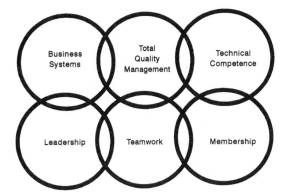

Figure 10.1   Areas contributing to organizational success.

when they are applied. Successful application is the key for results. Knowledge and skills have little or no impact on the organization if they are not applied successfully.

The organization facilitates the development of the right competencies, but each individual is personally responsible for developing them. Competencies help to define a blueprint leading to successful performance. Although competencies vary from organization to organization, the competencies required for excellence are well known. Some of the major individual competencies valued in a TQM organization are

- Ability to work in teams
- Customer orientation
- Ability to see systems view
- Ability to do high-quality work
- Ability to do a high quantity of good work
- Use of resources and time
- Communications
- Interpersonal relationships
- Conceptual skills
- Problem-solving skills
- Job knowledge
- Organization of work
- Personal incentive
- Ability to develop others
- Technical and professional competence

Learning process

A person can become competent in a variety of ways. Refer back to Fig. 3.6, to see what elements are essential for a successful continuous learning process, one in which competency is developed. The learning process starts with knowledge as the foundation. Knowledge is gained mainly through self-development and education. Skills are the next element of the learning process. They can be gained by training and experimentation. Next, the person needs to apply her or his knowledge and skills on the job. Facilitation and on-the-job training help this process. Finally, the learning process involves experience. Coaching and mentoring can be used to improve the learning process at this stage. All this learning takes place in the context of the organizational culture. A person may be competent in one organizational culture and not competent in another organizational culture. Figure 10.2 provides more specific methodologies for each of the development areas.

Figure 10.2   Learning process with methodologies.

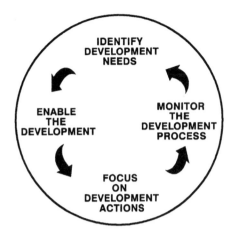

Figure 10.3 Development action process.

Development Action Process

Each individual needs to use a development action process to become competent within his or her own organizational culture. The development action process, as shown in Fig. 10.3, consists of the following:

1. Identify development needs.
2. Enable the development.
3. Focus on development actions.
4. Monitor the development process.
5. Regularly review progress on the individual action plan.

Identify development needs

Development needs come from the competencies of both the organization and the individual. These needs are identified through a variety of methods. The first step is to decide the specific competencies required for success in the organization. To do so the organization must perform a detailed analysis of present and future competencies. Next, the organization develops a valid instrument to assess the competency of the individuals in the organization. Each individual in the organization can use the assessment to determine where he or she currently stands. This self-assessment starts the development process. This tool assists the individual in the identification of individual development opportunities. In addition, you are advised to have others also complete assessment instruments to provide you with additional feedback. Self-assessment and feedback are both useful in identification of development needs.

Organizational assessment. The organizational assessment involves determination of current and future competencies for the organization. There are several ways to accomplish this activity. Some organizations seek assistance from one of the many organizations that specialize in this area. In addition, there are many standard instruments available to help in this task. Appendix A provides some examples for assessment of organizational success. For many organizations, a self-developed assessment is sufficient. The following action process assists in identifying competencies:

1. Identify a common view of the future "ideal" or "excellent" organization.

2. Determine a list of characteristics required to be successful in this organization. This activity can be done through brainstorming sessions, interviews, focus groups, etc.

3. Compile a list of potential competencies.

4. Select target competencies from the list.

5. Validate a list of competencies that would achieve the "ideal" organization.

6. Perform an assessment of each person in the whole organization using "ideal" competencies.

7. Finalize a list of competencies for the organization.

Self-assessment. The process of change begins with each individual. It is through the individual's mental outlook, personal behavior, and daily actions that change takes place in the daily routine of the workplace. While the organization can open up opportunities, the individual must take advantage of those opportunities for anything to happen. Individuals must seek out chances to grow and develop. They must use their new capabilities to improve the organization.

Each individual member of the organization is responsible for her or his specific evolution. Nothing will change unless there is awareness and understanding of one's potential. It requires the commitment of each working person, regardless of his or her role or level in the organization.

Self-assessment starts with an attitude. The person must welcome feedback and be willing to accept it with a healthy attitude to improve oneself. Self-assessment requires an attitude of openness so that feedback can be accepted from customers, peers, employees, and leaders.

Self-assessment answers the basic question, "How can I best improve myself to enhance my contribution to the success of the organization?" Answers to this question can be solicited formally through assessment. In addition, feedback from leaders, managers, peers, customers, and suppliers during daily activities provides regular feedback about competency.

Sample self-assessment. The following sample self-assessment includes the individual competencies that a particular organization determined were necessary for success in that specific organization.

Ability to work in teams

1. Do you prefer to spend time working on individual tasks or on team projects?
2. Do you spend time working on team projects?
3. As a member of a group, are you a team player?
4. Do you choose to be part of collective and shared effort?
5. Do you participate effectively in team meetings?

Customer orientation

1. Do you consider the "user" of your work as a "customer"?
2. Do you think about the satisfaction of the "customer" of your work?
3. Can you put yourself in the place of your customer?
4. Do customers' needs and expectations drive your work?
5. Do you concentrate on "doing the right thing right the first time" for your customer?
6. Do you ensure that your requirements are defined for your suppliers so that your customers are satisfied?

Ability to see systems view

1. Do you view the whole process in which you work to include the organizational system, total workflows, suppliers, and service to the ultimate customer?
2. Do you see the big-picture issues in the organization?
3. Do you design and create your work process to optimize the system?
4. Do you review your work process with your immediate customer?
5. Do you ensure that your work process adds value to the ultimate customer?
6. Do you see how your work process fits into the overall system?
7. Do you understand your role in the total organization?
8. Do you make suggestions or improvements in your work process in view of the total system?
9. Do you consider the total system rather than only your function?

Ability to do high-quality work

1. Do you value doing work right?
2. Do you set your own higher expectations for the work?
3. Do you take pride in your work?
4. Do you feel you own your work?
5. Do you solicit customers' views of what high-quality work is and adjust your standards accordingly?
6. Do you do your work according to specifications built from customers' needs?
7. Do you seek feedback on the quality of your work?
8. Do you have metrics of your own performance?

Ability to do a high quantity of good work

1. Do you produce outputs according to customers' demands?
2. How good is your turnaround, as customers see it?
3. Do you produce a high volume of work?
4. Are you responsive to customers' expectations of performance?

Use of resources and time

1. Do you use time and resources effectively and efficiently?
2. Are you aware of the cost of doing business or your job?
3. Do you consider the passing of time as an expenditure of money in the work?
4. Are you cost conscious?
5. Do you use technology, when appropriate, to assist you in managing tasks, time, and resources?
6. Do you contribute to the reorganization of work and priorities to optimize resources?
7. Do you plan, schedule, and control work to optimize workload?

Communications

1. Do you speak clearly?
2. Do you make effective presentations to groups?
3. Do you keep your communications to the point?
4. Do you communicate specifically to the audience?
5. Do you prepare clear and concise written communications?
6. Do you listen effectively?

7. Do you actively listen to your customer(s)?

8. Do you provide feedback when warranted?

9. Do you encourage give-and-take discussions?

10. Do you express yourself openly and honestly?

Interpersonal relationships

1. Do you interact successfully with a wide range of people?

2. Do you focus on the process and not the person?

3. Can you maintain control and composure in conflict situations?

4. Do you interact successfully with a wide range of people?

Conceptual skills

1. Do you think critically about the issues and look beyond superficial symptoms in order to discover underlying causes?

2. Do you have an action process in your mind about how things should go before you pursue a plan of action?

3. Do you have an understanding of the customers' needs and expectations so you can meet them?

Problem-solving skills

1. Do you readily identify critical problems?

2. Are you effective at identifying the "real" problem and addressing it?

3. Do you simplify problems in order to solve them?

4. Do you seek root-cause analysis?

5. Do you explore patterns and trends?

6. Do you gain consensus on issues and solutions before taking action?

7. Are you comfortable working plans and programs to solve problems?

8. Do you stay with implementation to see the problem through resolution?

9. Do you use a systematic approach to problem solving?

10. Do you develop plans with milestones for solutions?

11. Do you determine what to do, by whom, in what order, and when for actions?

Job knowledge

1. Do you keep current in your area of work?

2. Can you effectively translate your technical knowledge to guidelines for others?

3. Are you active in understanding the customer's job and technical expertise?

4. Do you pursue the latest technical knowledge in your field?

Organization of work

1. Do you set clear objectives?

2. Do you stay focused on objectives until they are complete?

3. Do you work several tasks at once?

4. Do you prioritize tasks?

5. Do you work one task at a time?

6. Do you use technology to assist you in organizing your work?

Personal initiative

1. Do you take the initiative to change processes and procedures that do not work?

2. Do you willingly take on jobs and tasks that are not part of your job?

3. Do you communicate your ideas for continuous improvement, even if they suggest more work for you?

4. Do you do whatever is necessary to get the work done?

5. Do you focus on accomplishments rather than activities?

6. Do you measure results by achievement rather than hours worked?

7. Do you enjoy the challenge of work?

8. Can you perform independently?

Ability to develop others

1. Do other people seek you out for advice?

2. Do you take enough interest in the growth of others to listen to them?

3. Do you act as a resource person?

4. Are you accessible to others?

5. Do you offer yourself to coach others?

6. Can you train others to do your job?

7. Do you recognize as part of your job the development of others?

8. Do you support and provide alternatives for development?

Technical and professional competence

1. Do peers seek your advice on technical issues?
2. Do you assess your own competence?
3. Do you seek feedback on your competence from others?
4. Do you take the opportunity for self-development and professional development activities?
5. Do you design your own measures for your technical effectiveness?

Enable the development

You can enable development though action planning. This plan ensures that specific actions are accomplished so that people in the organization become competent. The action plan should emphasize both short-term and long-term development needs. Short-term needs are categorized as development needs, and long-term needs are growth areas. These needs should be included in an individual development action plan. The action plan provides the what (actions), when (milestones), and how (education, training, coaching, mentoring, and on-the-job training). Figure 10.4 shows an individual development action plan worksheet.

Focus on development actions

The individual takes appropriate action, which helps build competence to improve that person's effectiveness. The individual must constantly be engaged in education, training, and coaching activities.

Education. Education targets the gaining of knowledge. Education provides the what. It supplies an understanding of ideas, concepts, theories, history, background, situations, and so on, and it instills the knowledge to intelligently select from many alternatives for a specific situation.

Education can be gained from a variety of sources, from formal methods such as seminars, lectures, and college courses, to more informal methods such as self-study, observation, and interviewing.

Training. Training begins the focus on application. It not only involves knowledge, but it also considers attitude and skills to perform specific tasks. Training is geared toward a specific need of the individual and the organization.

Training can be received from workshops, seminars, and on-the-job training.

Development Item	Development Method	Scheduled Date	Completion Date

Figure 10.4   Individual development action plan worksheet.

Formal training action process

1. Analysis
    a. Needs
    b. Performance
    c. Process
    d. Operations
    e. Tasks
2. Design
    a. Objectives
    b. Lessons
    c. Measurements
3. Development
    a. Lesson plans
    b. Training materials
4. Implementation
    a. Prepare for training
    b. Present training
    c. Administer training

5. Evaluation
   a. Review impact of training on business results

On-the-job training action process

Prepare—Provide knowledge by telling, reading, etc.

Demonstrate—Let trainee watch actual operation while instructor explains.

Practice—Let trainee do it while instructor watches and makes corrections.

Evaluate—Let trainee perform while instructor assesses performance.

Coaching. Coaching involves intensive application focused on specific results. Coaching is geared to the particular needs and situation of an individual or team. The coach provides analysis, planning, instruction, application, and evaluation.

Coaching action process. Coaching involves the continual use of the coaching process. For example, first the coach helps the top executive assess the organization. Second, she or he assists the top executive in the development of an action plan to achieve the specific VICTORY. During this step the contract with the coach is established. Third, the coach provides instruction on the process which, in many cases, can be given through self-instruction materials tailored to a particular situation. The coach, prior to application, provides a practice session to ensure that the top executive knows the specific process to be performed—that is, what needs to be done and how to do it. To be successful, the top executive needs to be able to apply the instruction to each specific situation. Fourth, the coach and top executive evaluate performance. The performance evaluation leads to a repeat of the cycle until VICTORY can be achieved without further coaching. The coaching process involves

- *Analysis*—assessing the situation
  - Performing assessments
  - Asking questions
  - Reviewing experience
- *Planning*—creating the game plan
  - Brainstorming ideas
  - Clarifying ideas
  - Deciding "critical" issues

- Finalizing an action plan
- *Instruction*—practicing the methodology, techniques, skills
  - Education
  - Training
  - Facilitating
  - Mentoring
- *Application*—playing the game
  - Performing actions from action plan
- *Evaluation*—reviewing the results
  - Reviewing content, process, and results
  - Discussing what went well and alternatives
  - Providing positive reinforcement and encouragement
  - Getting agreement on next steps

Facilitating. Facilitating involves assisting with the application of a specific process to make it easier to achieve desired outcomes. It helps the learning by doing. This skill is usually a team or group intervention which is geared to the particular needs, expectations, and situation of the organization.

Facilitators are important support people in a TQM organizational support structure. They help the team reach its mission without being team members. Through their assistance, they can develop team leaders and team members while those people are actually performing their tasks.

The facilitator acts as a process expert to a group or team. He or she also helps that group or team apply specific philosophy, principles, processes, methodology, tools, and techniques to achieve outcomes. The facilitator ensures that the group or team does the right things right.

Facilitator role. The facilitator role is to be that of a(n)

Educator	Provide specific knowledge
Trainer	Equip with specific skills
Motivator	Give encouragement, support, and recognition
Coach	Develop total capabilities of the team
Mediator	Promote a collaborative environment
Negotiator	Assist in finding win-win outcomes
Devil's advocate	Stimulate open communication
Public relations	Advertise team's successes

Story teller	Communicate team's culture
Confessor	Listen to team members
Cheerleader	Provide boost to induce progress
Supporter	Assist with obtaining resources

Facilitator contributions. The facilitator is a resource to use to help achieve success. She or he brings the following contributions to help the team:

- Knowledge of appropriate concepts, philosophy, principles, processes, methodologies, tools, and techniques
- Skill in letting the team perform, especially allowing the leader to lead
- Competency in effective team meetings
- Ability to keep the team focused
- Capability to apply the appropriate tools and techniques for training
- Aptitude in using the right intervention techniques
- Proficiency in coaching
- Ability to manage conflict
- Talent for communicating, especially active listening

Facilitating action process. Facilitating is useful for any situation involving a group of people working together. The use of a facilitator who is an expert in effective team meetings, team dynamics, interpersonal relations, improvement processes, tools, and techniques allows the team to concentrate on using their capabilities to achieve the desired outcomes without the additional worry of "how to" work as a team. The facilitator works the process; the team works the content. The facilitator guides the team through the process as painlessly and quickly as possible. The facilitator action process involves the following:

- Analysis
  Perform a review of the agenda before the meeting.
  Assess the team leader and team members' dynamics.
- Planning
  Generate alternatives.
  Evaluate alternatives.
  Develop plan to act.

- Taking action

  Perform the action considering the focus and relationships.

- Evaluation

  Check the results of the action.

  Provide feedback as necessary.

  Review meeting with team leader after the meeting.

Mentoring. Mentoring involves the provision of an experienced person to offer guidance and support. In TQM, team leaders and teams find that mentors are particularly useful, as they serve as management interface and support. As such, they need to have credibility and influence in the organization. They support the team leader by offering a listening ear. They also are useful to provide advice on handling issues that involve internal politics. Frequently, a mentor can assist the team in introducing and implementing a critical TQM initiative that is counter to the present organizational culture.

Mentors are important in individual development. Personal attention by an experienced person provides insight that leads to success.

Mentor's role. The mentor's role is

- Role model
- Coach
- Performance monitor
- Guidance counselor
- Political advisor
- Devil's advocate
- Mediator

Mentoring action process. The following action process promotes successful mentoring:

1. Define mentor criteria.
2. Select mentor and protégé.
3. Develop mentor and protégé relationship.
4. Systematize the process:
   - Assessment
   - Performance
   - Review and encourage
5. Evaluate mentor and protégé relationship.

Monitor the development process

Regularly review progress on the individual action plan to ensure that

- The individual's development progress is on track
- The individual's development supports the changing organizational views
- Encouragement and support are offered
- Recognition and rewards are provided
- Additional items for individual action plan are generated
- Individual performance is determined
- Self-assessment is reevaluated
- What is going well is recognized
- Lessons learned are compiled

Yearning for Success:
The Obligation of Leadership

If you want help from others, you must be willing to help yourself.

Leaders must have the intense desire to win. In addition, the people in the organization must believe enough in their leader to devote their energies toward her or his vision. This endeavor requires both leaders and followers to make the personal commitment necessary for success.

Leaders must commit to long-term support. They must be willing to make an investment of their personal time and the organization's resources. They must also understand that although some results will be quickly realized, permanent changes will take many years. Leaders' long-term support involves many actions. First, leaders must set the example by displaying expected behaviors day after day. Second, they must supply resources, including people, time, and money. Third, they must provide the direction, guidance, and support to the overall TQM effort. Fourth, leaders must deal with difficult issues. Fifth, they must think progressively. In summary, leaders must be active, highly visible participants in all aspects of the TQM process.

Followers must commit to a belief in their leaders and be willing to trust them. They must understand that doing their job in the foreseeable future will require them to have a different way of thinking. It will mean that they must go beyond "just doing the job." They will soon recognize that doing their new tasks will not be easy and there will be some hardships, mistakes, and frustration along the way. But they must also realize that the potential for personal and organizational benefits is great.

Members in the organization are indeed critical components of a successful TQM effort, but so are leaders who demonstrate a yearning for success by their active involvement, commitment, and support. Leaders must have the discipline to make this long-term commitment for the future of the organization; they must also thoroughly understand the TQM process. In addition, they must devote personal attention to the TQM philosophy and guiding principles to see that they are constantly and consistently applied throughout the entire organization. This chapter will discuss the various roles and responsibilities of leaders in the organization.

Leaders show yearning for success by

- Setting the example
- Inspiring others
- Dealing with difficult issues
- Thinking progressively
- Taking calculated risks
- Learning from the mistakes of yourselves and others

Leaders Guide Success by Setting the Example

> *Set the example and others will follow.*

As a leader, one of your most important roles is to set an example for others to follow. Chapter 3 outlines considerations for setting the example.

Leading by example action process

Leaders have an impact on the organization, either a positive or negative one. We all have an example of a leader's behavior having a positive or negative impact on the organization. The following action process will help in the development of leaders in your organization who inspire others to VICTORY.

1. Determine how the behaviors of leaders in your organization, especially top leaders, influence the behaviors of other leaders in that organization.
2. Define the kind of organizational environment you want to foster.
3. Specify actions that leaders could take to lead by example so that the desired organizational environment can eventually be established.
4. Develop an action plan.
5. Monitor and follow up on the plan and desired outcomes.

Leaders Inspire Others

> *Inspiration is better than perspiration.*

Besides setting an example that others can follow, leaders must inspire others to work toward a common purpose. They inspire others by involving them in the vision, which then becomes a view of the future shared by everybody in the organization. Leaders who can be trusted, have ethics, and display integrity induce others to common principles. Those who perform certain actions can produce desired inspirational results. Examples of such actions and results are

Raising standards stimulates the spirit to keep growing.

Pursuing constructive relationships motivates others to a positive work environment.

Inviting open and honest communications sparks creativity and innovation. Respecting other's points of view encourages higher performance.

Encouraging change, innovation, and risk taking results in wins.

Inspiring considerations

Involve others in the vision to make it shared.

Nurture trust, ethics, and integrity.

Set high standards to make things good, better, best.

Pursue constructive relationships.

Invite open and honest communications.

Respect others' points of view.

Encourage change, innovation, and risk taking.

Deal with Difficult Issues

> *Winners accept the challenge of difficult situations.*

Leaders learn to deal with difficult issues, a task which separates the winners from the losers. Winners see difficult issues as opportunities; losers see them as problems. Effectively facing difficult issues instills confidence in the organization to face the ever-changing environment.

---

1. **Define the difficult issue**. What is the "real" issue?

_____

_____

_____

2. **Envision the outcomes expected to be achieved**. How will you know the issue is resolved? What will it look like when the issue is resolved?

_____

_____

3. **Action plan**. How do you plan to deal with the issue? Include how you plan to use specific Team Leader module(s).

_____

_____

_____

_____

_____

4. **Lead the organization/team to resolve the issue**. What is the recommended solution to deal with the issue?

_____

_____

_____

Figure 11.1   Dealing with difficult issues worksheet.

Dealing with difficult issues action process

The following is the action process for dealing with difficult issues:

Define the difficult issue.

Envision the outcomes expected to be achieved.

Action plan—establish and implement one.

Lead to resolve the issue.

Figure 11.1 shows a worksheet for dealing with difficult issues.

Think Progressively

> *If you always think the same, you will always be the same.*

Leaders must think progressively for the organization to progress. Today, organizations must lead change. This effort requires the top leaders and all the leaders in the organization to be constantly thinking about the evolving environment and the changes required within the organization for it to be competitive.

Thinking progressively action process

1. Take time to regularly monitor the environment:
   - Decide what to monitor.
   - Develop an instrument to survey the climate.
   - Conduct a survey of the environment.
   - Analyze the results of the survey.
   - Determine organizational changes.

2. Have structured activities for systematic changes:
   - Decide what activities or interventions are appropriate.
   - Perform project management planning for changes.

3. Involve all key stakeholders:
   - Establish champions for specific change(s).
   - Make a case for change(s).
   - Form teams for actions.

4. Nurture the needed change(s):
   - Create an appropriate sense of urgency for change(s).
   - Provide recognition and rewards.

5. Know the progress of the change:
   - Establish a process to evaluate the impact of the change(s).
   - Celebrate accomplishments.
   - Communicate lessons learned.

Take Calculated Risks

> *Risk means opportunity.*

Calculated risk taking involves taking risks where possible consequences are well considered and evaluated with the potential rewards greater than acceptable and affordable losses. Calculated risk taking is systematic, reasonable, informed risk taking.

Most definitions of risk highlight the exposure to loss; risk also involves the potential for reward. There are two kinds of possible outcomes of risk:

1. Real reward or loss which leaves the decision maker better or worse than before the decision, that is, closer or farther from the objective.

2. Opportunity reward or loss which is an outcome that is more or less favorable than it could have been, that is, maybe another approach would have yielded a better or worse result.

Risk can be frightening, but usually when the risk is the greatest, the potential for reward is also the greatest. Many of us focus on the exposure to loss, so it is no wonder we usually decide to take the path of least resistance, choosing risk avoidance to fulfill a perception of safety and security. Many people choose to act to avoid failure as opposed to seeking success. Some people live in the past; others in constant hope for the future; still other simply live in the present by the motto "don't rock the boat." In the workplace, traditional management practices foster fear—leading to unwillingness to take risk. When this attitude persists over a long time, comfort-seeking actions become the habit.

Risk taking has many rewards. It offers the opportunity for organizational success and personal satisfaction. It creates breakthroughs, innovation, and invention. Risk-seeking behaviors lead to success.

Calculated risk-taking action process

There is no "best" technique for risk taking. This section provides a calculated risk-taking process. The risk taker(s) must determine the appropriate content of the process for each particular situation. This content depends on the event and the outcome, the probability of an event occurring, and the significance of the outcome. The risk-taking process is as follows:

Recognize the risk.

Investigate the risk issue.

Seek actions to manage the risk.

Keep track of progress toward achieving a plan.

Steps to calculated risk taking

The calculated risk-taking process involves using a systematic, disciplined four-phase approach to making decisions in high-risk situations. The four-phase approach involves a 10-step process as outlined below.

- Recognize the risk:

    1. Identify the risk situation.
    2. Review the mission and objectives as they relate to the risk situation.

- Investigate the risk issue:

    3. Define the risk issue(s) criteria.
    4. Analyze the risk issue(s).

- Seek actions to manage risk:

    5. Identify alternatives.
    6. Evaluate alternatives.
    7. Select a course of action.
    8. Gain support from stakeholders.
    9. Develop a plan of action.

- Keep track of progress toward achieving the plan:

    10. Track progress against the plan and adjust as necessary.

Process 1: Recognize the risk. Risk is present in some form and degree in most activities. It is critical to success to recognize and manage risk. The first phase of the calculated risk-taking process, recognizing the risk, involves two steps.

Step 1. Identify the risk situation. The identification of risk can come from many sources. Some potential sources of risk identification include

Output from technical, environmental, and relationship assessments

Results of understanding the process

Analysis of performance measurements

Evaluation of plans

Views of people

Indication of a problem

Once a potential risk situation is identified, the next activity involves determining if you in fact have a risk situation. A simple decision tree will help you with this activity.

First, state the risk situation. Determine if the situation is certain or uncertain (risky). If it is certain, there is no risk. Stop the process. If there is any uncertainty, continue to the next item.

Second, estimate the potential reward/loss ratio. Ask: Are there possible significant rewards and losses in the risk situation? "Significant" varies by the situation, so it must be determined by the decision maker(s). This statistical determination can be made through use of highly objective methods, highly subjective methods, or a combination of objective and subjective methods. Normally, subjective methods are used during this step. If there are no possible significant rewards or losses, stop the process. If there is possibility of a significant reward or loss, go to step 2.

Step 2. Review mission/objectives as they relate to risk.   Once you recognize a potential risk situation, you need to revisit the organization's and team's mission and objectives to understand the really important elements in regard to the risk situation.

Once more, there is an evaluation of the risk situation. Is the risk situation acceptable or unacceptable in terms of achievement of the mission/objectives? Again, what is acceptable and unacceptable will vary by the situation. The decision maker(s) determine the criteria. As before, this statistical determination can be made through use of highly objective methods, highly subjective methods, or a combination of objective and subjective methods. Normally, subjective criteria is used during this step. If the risk situation is acceptable, stop the process. If the risk situation is unacceptable, continue to step 3.

Process 2: Investigate the risk issue.   The second phase of the calculated risk-taking process—investigating to find the risk issue—involves two more steps.

Step 3. Define the risk issue(s) criteria.   Once you determine, that there is a risk worth investigating, develop a consistent scheme for rating risk. Make it quantitative with qualitative backup. The risk issue must be described and documented sufficiently to provide some criteria for prioritization of risks for the analysis step. For instance, in a program there could be a risk rating for cost, schedule, performance, and/or some other measurable factor. Again, this statistical determination can be highly objective, totally subjective, or a combination of objective and subjective. Normally, subjective criteria are used during this step. Heavy mathematical treatment is not necessary at this step. For example, the following method could be used:

A rating scheme provides a framework for eliminating some of the ambiguity associated with many people looking at a risk situation. The rating system should be as simple as possible, such as high, medium, and low.

Expert interviews provide information from technical experts which can be used to set up a gauge for analyzing the risk issues.

Analogy comparisons provide criterion from other similar existing or past programs.

Step 4. Analyze the risk issue.   Risk analysis involves an examination of the change in consequences caused by changes in risk input variables. Sensitivity and "what-if" analyses are examples of activities that should take place during analysis of the risk situation. During this step, use an analysis tool designed to meet your specific objectives.

Depending on your objective, the results of analysis can be examined in terms of the following:

Cost/schedule/performance/customer satisfaction

System/subsystem

Funding profiles

Criticality

Consistency with analogous systems

What-if scenarios

Task risk facets

Relationship factors

During this step, the "as is" reward/loss ratio is quantified. In addition, goals are set to specify outcomes. The "as is" reward/loss ratio is evaluated in regards to meeting goals.

Process 3: Seek actions to manage risk.   The third phase of the calculated risk-taking process—seeking actions to manage risk—involves five steps.

Step 5. Identify alternatives.   For each risk-taking situation, the team must decide a risk strategy to drive actions. Calculated risk-taking strategy includes several options as follows:

Avoid risk

Assume risk

Control risk

Transfer risk

Share risk

Study risk

Alternative actions are generated based on the calculated risk-taking strategy.

During this step first look at the process causing the risk situation to determine if there is something obvious that can be done with the process to make the risk acceptable. Ask: Can the process be changed to make the risk situation acceptable? Are there any non-value-added tasks? Can the process be simplified? Can activities be combined? Is there another method available? Can people be trained? Would written procedures help? Will the alternative action produce your objectives?

In addition to the obvious alternatives, the team should explore "breakthrough" alternatives. For example, challenge the old "rules." Ask: Does the process need to be reengineered? Can information technology provide a better process? Are as few people as possible involved in the process? Does the process eliminate as many non-value-added activities as possible?

Step 6. Evaluate alternatives.   During this step, evaluate alternatives based on the selection criteria developed during process 2, step 3. Ask: Which of the alternatives will get you closest to your objectives?

Step 7. Select a course of action.   In this step, the team decides on the "best" course of action which does the following:

is consistent with mission/objectives

meets risk criteria and strategy

uses all relevant, available information

It is important to use consensus decision making to make this decision.

Step 8. Gain support from stakeholders.   In order to implement the selected course of action, it is critical to get the support of key stakeholders. The stakeholders can help ensure that all consequences of a decision are weighed properly. This step may involve the preparation and conducting of a presentation.

Step 9. Develop a plan of action.   One essential step in ensuring that the risk is managed properly is the development of a plan of action.

Process 4: Keep track of progress.   The last process is an ongoing activity of continuous evaluation of progress. The step may lead to a return to step 1 and a looping through the calculated risk process again and again.

Step 10. Track progress against the plan and adjust.   During this step, the progress and success are measured against the plan to ensure that the risk is managed properly.

Learn from Mistakes

> *Failures breed successes.*

Organizations and people learn more from failure than from success; sometimes you can learn a lot from the mistakes of others. Success in TQM cannot be guaranteed, so do not be discouraged by some failures—but do everything possible to avoid fatal errors. The following are some of the lessons learned from the implemention of TQM.

Lesson 1: Expecting results too quickly.   Implementation of TQM produces some results immediately. However, the big payoff requires many years of commitment and support. The greatest benefits of TQM come when it is institutionalized, which may take many years to achieve.

Lesson 2: Copying from other organizations.   There are benefits from learning from the successes and failures of other organizations. However, the real success from TQM comes from the persistent application of the TQM philosophy and guiding principles in each organization's specific environment. VICTORY is different for each organization. It cannot be achieved by simply copying others.

Lesson 3: Starting with insufficient resources to create a TQM environment.   VICTORY requires the full support and commitment of the entire organization. Such resources as funds, manpower, facilities, training, support structure, and in some cases technology are needed.

Lesson 4: Thinking training is all that is required.   Training and education are important elements of VICTORY, but training alone will not give success. All the elements of VICTORY are needed for the TQM effort to be successful.

Lesson 5: Setting goals that are not attainable. Goals are essential to focus the organization on VICTORY. People strive to achieve challenging goals, but no one wants to pursue unrealistic ones. Frequently, organizations set goals beyond their reach at first. The organization should set realistic goals to build on their successes. Remember, small successes repeated over and over build to big victories.

Lesson 6: Trying to solve the biggest problem all at once. The TQM process is based on the achievement of many small successes over time. Many of the problems of organizations evolved over many years; they cannot all be solved at once. Although it is important to focus improvement efforts on critical issues, they should be made little by little until the major issues are resolved. Trying to solve the biggest problem all at once will only result in frustration and failure.

Lesson 7: Running TQM like a program. TQM is a way of life; it is not a program. TQM requires many changes in behavior that cannot be demanded. The goal of TQM is to institutionalize the philosophy and guiding principles into the organization. This goal can only be accomplished by continuous actions focused on reinforcing TQM behaviors.

Lesson 8: Implementing only some of the elements of VICTORY. All of the elements of VICTORY are necessary. Some organizations pick and choose certain elements, expecting results. This implementation will not work.

Lesson 9: Lacking integrity, ethics, and trust. Integrity, ethics, and trust are the underlying foundation of TQM. These qualities must be ingrained into the organization environment for VICTORY to be achieved.

Lesson 10: Lacking a clear vision purpose or purpose that cannot be made real. A clear vision that can be made real by the organization is of primary importance for VICTORY. Without a mission, the organization cannot start toward success. Also, the vision must be understood by the people who must make it happen. Everyone in the organization must see how they contribute to VICTORY.

Lesson 11: Lacking an overall plan. The creation and maintenance of TQM requires an overall plan. TQM does not just happen. It needs a systematic, integrated, consistent, organizationwide approach, which can only be achieved through complete planning.

Lesson 12: Paying lip service to improvement efforts. Total Quality Management must be a way of life to achieve VICTORY. More than words is required; action is needed to ensure the necessary TQM environment with all the elements of VICTORY focused on customer satisfaction.

Lesson 13: Practicing policies and procedures that do not support the TQM environment. All the policies and procedures in the organization must reinforce the TQM environment to ensure the TQM way of life. For instance, compensation policies should reward TQM behavior. Procedures should allow people ownership of their work.

Lesson 14: Failing to communicate successes. TQM spreads by word of mouth. TQM success should be constantly visible to everyone in the organization. When it is, the organization can build on successes for the future.

Lesson 15: Preaching one thing but doing another. TQM can only be established and maintained by the actions of the leadership. Leaders must consistently display the behaviors expected in a TQM environment.

Lesson 16: Failing to provide timely training. Sometimes TQM requires training to accomplish the improvement effort. Training should be geared to the specific improvement effort. All personnel on the team should go through the training together for each specific improvement effort. In addition, the training must be given to provide the skills for the improvement effort. These skills must be given just in time to accomplish necessary actions.

Lesson 17: Thinking once trained, always trained. Training must be continuously pursued.

Lesson 18: Failing to train, not simply educate, top leadership in TQM. Top leadership must thoroughly understand the TQM philosophy and guiding principles, and their application, as well as the continuous improvement system and tools and techniques before attempting to start organizationwide training for other members of the organization.

Lesson 19: Feeling you cannot do anything; it is not under your control. You can do many improvements within your organization, department, function, section, team, and yourself when you have control. Do whatever you can do; fix what you can. Remember, success breeds success. Your little improvements will lead to other little improvements, which will start others making improvements.

Lesson 20: Failing to communicate the meaning of TQM ownership. People require an understanding of the meaning of ownership in their organization. Frequently, management suddenly announces that all people in the organization now have ownership of their work. Most people have no idea what this means. Ownership must be defined by the amount of responsibility and authority given to the people. An understanding of meaning of ownership within the specific organization is essential to VICTORY.

Lesson 21: Implementing continuous improvement in only one area. Many organizations focus their improvement effort on one or two areas without involvement of other essential functions. For instance, the improvement effort typically starts in only the manufacturing, engineering, or human resources areas. TQM requires the involvement of all areas in the organization.

Lesson 22: Failing to balance short-term goals with long-term objectives. TQM requires a long-term perspective. Strive for short-term success focused on the long-term future of the organization. Many organizations are geared to the short-term gains or profits of the organization.

Lesson 23: Thinking technology will do it without people. Technology and people must be balanced in a TQM organization. Although there are definite advantages to technology, people adding value is a primary principle of TQM. People are the most important resource.

Lesson 24: Failing to listen. Listening is key to TQM. This is one of the major lessons learned. VICTORY requires listening to people in the organization, suppliers, and customers.

The Last Word to All Leaders

> *Optimize for VICTORY.*

The last word for success in any organization and especially in Total Quality Management is optimize. Too many leaders are looking for the one "right" answer; in many cases the only response to the question of whether the answer is *the* right one is "possibly." Optimization requires your personal long-term commitment, support, and active involvement. You are responsible for the ultimate success in your organization. You cannot defer your organization's success to others. You must optimize for your specific organization's

- Resources
- Life-cycle costs
- People and technology
- Process and content

- Results and relationships
- Supplier partnerships and customer focus
- Short-term and long-term orientation
- Revenue and cost
- External assistance and internal capability

There is no magic, no quick fix, no instant formula, no panacea technological advancement for success. Organizational success requires hard work, personal commitment, and perseverance. It also heavily depends on people. To achieve success for your organization you as a leader must at least

Focus on your customers

Lead the organization toward VICTORY

Visualize a common purpose for the organization

Involve everyone and everything

Continuously improve product, processes, and people

Train, educate, coach, facilitate, mentor, etc.

Own, and foster empowerment

Reward and recognize appropriate behavior, actions, and results

Yearn for success

Assessments

Assessment should be the starting point for any organization's improvement effort. This appendix provides information to formulate self-assessments for any organization to start its journey to continuous improvement. Simple assessment questionnaires are provided for the following areas:

- VICTORY assessment
- Teamwork development

Introduction

One of the critical elements of Total Quality Management is the constant assessment of the many factors contributing to success. The simple, easy-to-understand VICTORY model presented in this book is sufficient for most organizations today. This appendix and other assessment instruments within the book furnish abundant information to help you in formulating your own specific assessment instruments for continuous improvement of your organization to achieve its desired results.

For further information for assessments refer to the following:

1. The Malcolm Baldrige National Quality Award Criteria provides an excellent source for performing a top-level assessment. Copies of the complete 1996 Award Criteria can be obtained from:

   Malcolm Baldrige National Quality Award
   National Institute of Standards and Technology
   Route 270 and Quince Orchard Road

Administration Building, Room A537
Gaithersburg, MD 20899
http://www.nist.gov/

2. Another excellent assessment instrument developed for defense organizations but applicable for many organizations is the *Quality
   and Productivity Self-Assessment Guide.* For further information
   about the guide, contact the Defense Productivity Program Office.
   This guide not only is an assessment but it also contains specific
   suggestions and actions for improvement.

3. The ANSI/ASQC Q9000-94 Standards can be obtained from

   American Society for Quality Control
   310 West Wisconsin Avenue
   Milwaukee, Wisconsin 53203

When you are developing your particular assessment instrument,
ensure that the assessment is accurate, valid, reliable, understandable, and usable in your specific organization.

- The assessment is accurate if it provides correct and valuable feedback.

- A valid assessment gives information that is relevant and meaningful to your specific organization.

- A reliable assessment would provide the same results on successive
  trials.

- The assessment is understandable if it can be comprehended by the
  person taking the assessment and its results are clear to everyone.

- A usable assessment means it leads to meaningful action within
  your specific organization.

The format and rating scale of the assessment can take many
forms. The following are some examples for your consideration.

Example 1 *Instructions:* Please provide an accurate self-assessment. Circle the
appropriate number as it applies to you. The rating scale is as follows:

1 = Never

2 = Seldom

3 = Sometimes

4 = Usually

5 = Always

You tell others how you are feeling.

Never	Seldom	Sometimes	Usually	Always
1	2	3	4	5

If you never tell others how you are feeling, you would circle the 1. If you always tell others how you are feeling, you would circle 5. If you do not fit any of these extremes, you would circle the 2, 3, or 4, depending on your normal behavior.

Example 2 *Instructions:* Please provide an accurate assessment of your team. Check the appropriate box as it applies to your current team. Check the first box, "Not applicable," if you feel the item does not apply to your team. Check "Requires *no* improvement," if you know the item requires no further development at this time. This is an item the team has already accomplished or that the team is performing satisfactorily. Check "Requires improvement," if you feel your team requires improvement in this area. This could be an item that the team does not currently perform or which the team leader or the team feels requires improvement. If possible, please have each member of the team complete this team assessment.

NOT APPLICABLE _____ REQUIRES *NO* IMPROVEMENT _____
REQUIRES IMPROVEMENT _____

Example 3 *Instructions:* Please provide an accurate organizational assessment. Circle the appropriate number as it applies to your organization. The rating scale is as follows:

$$0 = \text{Don't know}$$

$$1 = \text{Highly inaccurate}$$

$$2 = \text{inaccurate}$$

$$3 = \text{Somewhat accurate}$$

$$4 = \text{accurate}$$

$$5 = \text{Highly accurate}$$

VICTORY Assessment

The purpose of this assessment is for you to accurately self-analyze your organization's environment. Please be open and honest in all answers. This assessment is held in strict confidence. You do not need to put your name anywhere on the survey.

Instructions: Please provide an accurate assessment of your organization. Circle the appropriate number as it applies to your organization. The rating scale is as follows:

1 = Never

2 = Seldom

3 = Sometimes

4 = Usually

5 = Always

Example   You tell others how you are feeling.

Never	Seldom	Sometimes	Usually	Always
1	2	3	④	5

The VICTORY assessment is based on the elements required for a winning organization. All the elements of VICTORY are necessary for survival today and success in the future. Leadership creates and maintains the VICTORY environment that focuses on the customer.

Leadership

1. Top management sets the example for the entire organization.

Never	Seldom	Sometimes	Usually	Always
1	2	3	4	5

2. Top management provides the guidance, means, and encouragement for all workers to follow for success.

Never	Seldom	Sometimes	Usually	Always
1	2	3	4	5

3. People in the organization look to top management for leadership.

Never	Seldom	Sometimes	Usually	Always
1	2	3	4	5

4. There are other people in the organization besides top management providing leadership when appropriate.

Never	Seldom	Sometimes	Usually	Always
1	2	3	4	5

How would you rate the organization's leadership?
Please grade according to the following:
"A" excellent, "B" good, "C" satisfactory, "D" less than satisfactory, "F" failing.

Grade _____

Vision

5. Top management has a vision of where the organization should go.

Never	Seldom	Sometimes	Usually	Always
1	2	3	4	5

6. The focus of the organization is communicated by top management throughout the organization.

Never	Seldom	Sometimes	Usually	Always
1	2	3	4	5

7. People in the organization can explain the purpose of the organization to others.

Never	Seldom	Sometimes	Usually	Always
1	2	3	4	5

How would you rate the organization's vision?
Please grade according to the following:
"A" excellent, "B" good, "C" satisfactory, "D" less than satisfactory, "F" failing.

Grade _____

Involvement

8. Management provides a work environment where everyone is focused on customer satisfaction.

Never	Seldom	Sometimes	Usually	Always
1	2	3	4	5

9. The work environment allows everyone to perform to the best of his or her abilities.

Never	Seldom	Sometimes	Usually	Always
1	2	3	4	5

10. People in the organization feel as if they really are the most important resource.

Never	Seldom	Sometimes	Usually	Always
1	2	3	4	5

11. There is trust between people in the organization.

Never	Seldom	Sometimes	Usually	Always
1	2	3	4	5

12. People in the organization are encouraged to be creative and innovative in their work areas.

Never	Seldom	Sometimes	Usually	Always
1	2	3	4	5

13. People in the organization participate in problem-solving activities affecting their work.

Never	Seldom	Sometimes	Usually	Always
1	2	3	4	5

14. Employees besides management are involved in decision making.

Never	Seldom	Sometimes	Usually	Always
1	2	3	4	5

15. Suppliers are an integral part of organizational planning.

Never	Seldom	Sometimes	Usually	Always
1	2	3	4	5

16. Customer inputs are solicited for most product-related efforts.

Never	Seldom	Sometimes	Usually	Always
1	2	3	4	5

17. Organizational systems (human resources, quality, policies, procedures, etc.) support the organization's objectives.

Never	Seldom	Sometimes	Usually	Always
1	2	3	4	5

18. Information is shared with all people in the organization requiring the information for job performance.

Never	Seldom	Sometimes	Usually	Always
1	2	3	4	5

How would you rate the organization's involvement activities?
Please grade according to the following:
"A" excellent, "B" good, "C" satisfactory, "D" less than satisfactory, "F" failing.

Grade _____

Continuous improvement

19. Continuous improvement of all systems and processes is fostered throughout the organization.

Never	Seldom	Sometimes	Usually	Always
1	2	3	4	5

20. A common methodology rather than fire fighting is the main problem-solving approach in the organization.

Never	Seldom	Sometimes	Usually	Always
1	2	3	4	5

21. The organization stresses prevention of errors rather than inspection as the major means to improve quality.

Never	Seldom	Sometimes	Usually	Always
1	2	3	4	5

22. Decisions are based on facts using quantitative methods, not opinions.

Never	Seldom	Sometimes	Usually	Always
1	2	3	4	5

23. People in the organization view their process (what they do) as a small business with suppliers and customers.

Never	Seldom	Sometimes	Usually	Always
1	2	3	4	5

How would you rate the organization's continuous improvement efforts?
Please grade according to the following:
"A" excellent, "B" good, "C" satisfactory, "D" less than satisfactory, "F" failing.

Grade _____

Training and education

24. The organization knows the attitudes, knowledge, and skills necessary for each person to be competent at his or her specific job.

Never	Seldom	Sometimes	Usually	Always
1	2	3	4	5

25. The organization provides training or education to help the people in the organization become competent at their current jobs.

Never	Seldom	Sometimes	Usually	Always
1	2	3	4	5

26. The organization provides training or education to help the people in the organization develop for advancement to other jobs.

Never	Seldom	Sometimes	Usually	Always
1	2	3	4	5

How would you rate the organization's training and education endeavors?
Please grade according to the following:
"A" excellent, "B" good, "C" satisfactory, "D" less than satisfactory, "F" failing.

Grade _____

Ownership

27. Everyone in the organization has ownership of his or her work.

Never	Seldom	Sometimes	Usually	Always
1	2	3	4	5

28. Everyone has the authority, responsibility, and resources to perform work at the best of her or his abilities.

Never	Seldom	Sometimes	Usually	Always
1	2	3	4	5

29. People in the organization take pride in their workmanship.

Never	Seldom	Sometimes	Usually	Always
1	2	3	4	5

30. People in the organization will give extra to fix a problem or improve the process.

Never	Seldom	Sometimes	Usually	Always
1	2	3	4	5

How would you rate the organization's efforts to instill ownership?
Please grade according to the following:
"A" excellent, "B" good, "C" satisfactory, "D" less than satisfactory, "F" failing.

Grade _____

Rewards and recognition

31. The reward and recognition system is fair.

Never	Seldom	Sometimes	Usually	Always
1	2	3	4	5

32. The reward and recognition system fosters the appropriate behavior of all employees.

Never	Seldom	Sometimes	Usually	Always
1	2	3	4	5

33. The reward and recognition system motivates people to strive for the overall success of the organization.

Never	Seldom	Sometimes	Usually	Always
1	2	3	4	5

34. There is constant motivation through the reward and recognition system.

Never	Seldom	Sometimes	Usually	Always
1	2	3	4	5

How would you rate the organization's reward and recognition system? Please grade according to the following:
"A" excellent, "B" good, "C" satisfactory, "D" less than satisfactory, "F" failing.

Grade _____

Yearning for success

35. Top management sets the example for change in the organization.

Never	Seldom	Sometimes	Usually	Always
1	2	3	4	5

36. Top management provides active support for change in the organization.

Never	Seldom	Sometimes	Usually	Always
1	2	3	4	5

37. Top management has the majority of the people in the organization's support for change.

Never	Seldom	Sometimes	Usually	Always
1	2	3	4	5

38. Most people in the organization feel a sense of urgency for making changes in the organization.

Never   Seldom   Sometimes   Usually   Always
  1   2   3   4   5

39. The organization operates in accordance with the highest ethical standards.

Never   Seldom   Sometimes   Usually   Always
  1   2   3   4   5

How would you rate the organization's yearning for results, full commitment, and willingness to support improvement efforts for many years?
Please grade according to the following:
"A" excellent, "B" good, "C" satisfactory, "D" less than satisfactory, "F" failing.

Grade _____

Focus on customers

40. Customers are the focus of most organizational efforts.

Never   Seldom   Sometimes   Usually   Always
  1   2   3   4   5

41. Everyone in the organization views other people in the organization as internal suppliers or customers.

Never   Seldom   Sometimes   Usually   Always
  1   2   3   4   5

42. Everyone in the organization knows their internal customer(s).

Never   Seldom   Sometimes   Usually   Always
  1   2   3   4   5

43. Everyone knows exactly what satisfies their internal customer(s).

Never   Seldom   Sometimes   Usually   Always
  1   2   3   4   5

44. Everyone in the organization knows how what they do impacts the satisfaction of the ultimate customer.

Never   Seldom   Sometimes   Usually   Always
  1   2   3   4   5

45. Everyone in the organization can describe the ultimate customer and what the organization does for the customer.

Never   Seldom   Sometimes   Usually   Always
  1   2   3   4   5

How would you rate the organization's focus on customer's? Please grade according to the following: "A" excellent, "B" good, "C" satisfactory, "D" less than satisfactory, "F" failing.

Grade _____

Team Assessment: Teamwork

The purpose of this assessment is for the team leader and/or team members to perform a self-analysis of their teamwork. This survey forms the focus for team development.

This section assists you with the assessment of teamwork within your team. The analysis is based on the characteristics of successful teams and applies to the complete team.

Trust

- Is the level of trust among team members sufficient to allow open and honest communication without tension?
- Can team members promote an innovative or creative idea?
- Do team members tell one another how they are feeling?
- Does the team share as much information as possible within the team?
- Does the team share information with outside teams, support, suppliers, and customers?
- Do all team members make an input into the decision-making process?
- Do team members keep team integrity and confidences?
- Do team member freely admit mistakes?
- Do team members avoid blaming, scapegoating, and fault-finding?
- Do team members demonstrate respect for one another's opinion?
- Do team members demonstrate respect for other peoples' opinion?
- Do team members directly confront issues with people rather than avoid or go around them?
- Do team members credit the proper individual(s)?
- Do team members always maintain everyone's self-esteem?

Effective communication, especially listening

- Does everyone have a chance to express his or her ideas?
- Do team members actively listen to one another?
- Is the message communicated clarified as much as possible?
- Do team members observe body language?
- Are team members' points made short and simple?
- Do team members take time to understand others' points of view?
- Are others' feelings considered when team members are communicating?
- Do team members involve themselves in the message being conveyed?
- Do team members generally comprehend most messages being communicated?
- Do team members pay attention to the message of others?
- Do team members talk judiciously?
- Is listening emphasized by all team members?
- Do team members listen actively while others convey their message?
- Is summarizing and paraphrasing used frequently to develop understanding?
- Do team members empathize with others' views?
- Do team members nurture active listening skills?
- Does the team foster an environment conducive to sharing feedback?
- Does the team encourage feedback as a matter of routine?
- Does the team have guidelines for providing feedback?
- Are all unclear communications openly discussed by the team?
- Is direct feedback given team members when appropriate?
- Do team members ask questions to get feedback?
- Do team members consider the "real" feelings of other team members when giving feedback?
- Is feedback focused on the issue and not made personal?

Attitude, positive "can-do"

- Do team members maintain a positive, "can-do" approach in all team activities?
- Do team members see opportunities even in negative situations?

- Do team members display a willingness to take risks?
- Do all team members maintain positive outlook even when faced with adversity during the second stage of team development?
- Do team members work toward a small success to overcome negative attitudes?
- Does the "team" explore the "root" causes of any negative attitude?

Motivation

- Are team members actively participating?
- Do team members demonstrate the self-confidence and self-esteem to actively participate on the team?
- Do team members direct all their energy toward the team's mission?
- Does everyone on the team truly believe that the team goal is shared?
- Are individual contributions by team members valued by other team members?
- Is the satisfaction of the needs of individual team members considered during team activities?
- Does everyone in the team know the team's rewards and recognition system?
- Does the team's reward and recognition system recognize intrinsic rewards such as a feeling of accomplishment, opportunity for personal growth, improvement of self-esteem, and sense of belonging?
- Does the team's reward and recognition system provide fair compensation, opportunity for advancement, and competitive benefits?
- Does the team's reward and recognition system credit actual performance?
- Do team members provide praise when it is deserved?
- Does the organization empower the team to provide recognition and rewards outside the normal system for extraordinary performance?
- Does the team celebrate successes together?
- Does the team allow fun?

"We" mentality

- Do team members demonstrate a togetherness in words and actions?
- Do all team members feel a sense of belonging to the team?

- Are important decisions based on consensus?
- Do team members cooperate rather than compete?
- Do team members orient toward the mission rather than the person?
- Do team members avoid making issues personal?
- Does the team focus on negotiating win-win solutions?
- Does the team take an organizationwide perspective?
- Does the team recognize conflict as natural?
- Do team members recognize the limits of arguing?
- Do team members empathize with others' views?
- Are all team members viewed as equal regardless of perceived status differences?
- Does the team examine all sides of every issue?
- Does the team support constructive relationships?
- Does the team recognize the strengths of individual differences?
- Does the team use individual differences as an advantage to achieve the mission?
- Does the team know too much agreement can be negative?
- Does the team appoint a "devils advocate" when everyone agrees too readily?
- Do team members maintain team integrity?
- Do team members avoid gossiping, faultfinding, blaming, and back stabbing?
- Does the team recognize that every team goes through four stages of team development?
- Does the team know their current stage of team development?

Ownership

- Do team members take the initiative to solve problems and/or improve their processes as a natural course of action?
- Do team members demonstrate constructive team behaviors to set the example?
- Are each team member's roles and responsibilities defined?
- Does each team member know her or his roles and responsibilities?
- Does each team member know the roles and responsibilities of all the other team members?

- Are individual contributions recognized by the team?
- Do team members take pride in the accomplishments of the team?
- Do all team members have the authority, responsibility, and resources needed to perform?
- Do all team members view themselves as owners?
- Do all team members understand the nature of their process with inputs from suppliers and outputs to customers?
- Do all team members know their suppliers?
- Do all team members know all their customers?
- Do all team members know all their customers' needs and expectations?
- Do all team members see themselves as suppliers of deliverables to others?
- Do all team members view their team as providing added value to a deliverable?
- Do all team members do whatever it takes personally to satisfy a customer?
- Does the team seek partnerships with suppliers?
- Does the team have an ongoing relationship with customers?
- Do all team members consider themselves marketers for the team's deliverable?
- Do team members take pride in the accomplishments of the team?
- Are team members empowered to perform and make improvements as necessary?
- Does the team's organization foster a constant learning environment?
- Are all team members technically competent?
- Can all team members describe and diagram their process?
- Do team members understand the business aspects of their process—for example, budgets, plans, competition, and return on investment?
- Does the team stress optimum life-cycle costs?
- Do team members concentrate on prevention of defects and designing in quality?
- Does the team use a disciplined approach to problem solving and process improvement?

- If the team's deliverable is to achieve total customer satisfaction, do they know through metrics to attain that goal?
- Does the team know the relationship of inputs to the output of their process?
- Does the team know their critical issues or opportunities?
- Does the team use data statistical analysis to analyze critical issues or opportunities?
- Does the team process improvements on critical issues and opportunities?

Respect, consideration of others

- Do team members respect individual differences?
- Are people's differences managed by the team to its advantage?
- Are constructive relationships being developed and maintained?
- Do team members feel that their individual self-worth is important?
- Do all team members treat others as they would want to be treated?
- Does the team seek outside assistance from others as necessary?
- Does the team coordinate activities with other teams and functions?
- Does the team actively seek to develop long-term relationships with customers, suppliers, other teams, and functional areas?
- Does the team have the respect of others in the organization?
- Does the team attend to the others' viewpoints?

Keeping focused

- Does the team know exactly what it needs to accomplish to be successful?
- Does the team have a clear focus?
- Does the team remain targeted on its focus?
- Does the team focus on the situation, issue, or behavior and not make it personal?
- Does the team have a vision of where they want to go in the future?
- Does the team have a common reason for action?
- Is the focus oriented to specific customer expectations?
- Does the focus target excellence?
- Is the vision set by the leaders in the organization?

- Do the leaders in the organization set the example for the focus?
- Is the vision communicated so everyone understands?
- Are sufficient resources committed to pursue the vision?
- Does the team have a mission statement?
- Is the mission statement customer-driven?
- Is the mission understood, clear, achievable?
- Does the team have consensus on the mission statement?
- Does the mission statement provide the purpose for the team?
- Does the mission statement set the common direction for the team?
- Does the mission statement set the expected results?
- Does the mission statement involve all team members?
- Did the team identify goals after a thorough analysis?
- Are the goals geared to a specific result?
- Can the goals be observed by measurement or metrics?
- Can the goals be achieved by the team?
- Do the goals provide a challenge for the team?
- Are the goals limited to a specific time period?
- Are all goals set by the team?

The following assessment applies to a quality improvement team.

- Was the mission originated by the manager or top leader?
- Was the mission negotiated and clarified by the team?
- Does the mission include the magnitude of improvement expected?
- Does the mission include the beginning process or perceived problem?
- Does the mission state the boundaries for the team?
- Does the mission state the authority for the team?
- Does the mission identify resources?

References

Amsden, Robert T., Howard E. Butler, and Davida M. Amsden. *Statistical Process Control, Simplified.* White Plains, NY, Quality Resources, 1989.

ANSI/ASQC Q9001-1994, *Quality Systems—Model for quality assurance in design/development, production, installation and servicing.*

ANSI/ASQC Q9002-1994, *Quality Systems—Model for quality assurance in production and installation.*

ANSI/ASQC Q9003-1994, *Quality Systems—Model for quality assurance in final inspection and test.*

Aubrey, Charles A., II, and Patricia K. Felkins. *Teamwork: Involving People in Quality and Productivity Improvement.* Milwaukee, Quality Press, 1988.

Barkley, Bruce T., and James H. Saylor. *Customer-Driven Project Management.* New York, McGraw-Hill, Inc., 1994.

Benedetto, Richard F., and Beverly Jones Benedetto. *Management Concepts for the 90's: Matrix and Project Management.* Dubuque, Kendall/Hunt, 1989.

Berry, Thomas H. *Managing the Total Quality Transformation.* New York, McGraw-Hill, Inc., 1991.

Block, Peter. *The Empowered Manager: Positive Political Skills at Work.* San Francisco, Jossey-Bass, 1991.

Brassard, Michael. *The Memory Jogger Plus.* Methuen, MA, GOAL/QPC, 1989.

Camp, Robert C. *Benchmarking.* Milwaukee, ASQC Quality Press, 1989.

Ciampa, Dan. *Total Quality: A User's Guide for Implementation.* Reading, MA, Addison-Wesley, 1992.

Cleland, David I. *Project Management: Strategic Design and Implementation.* Blue Ridge Summit, PA, TAB Books, Inc., 1990.

Creech, Bill. *The Five Pillars of TQM.* New York, Truman Talley Books/Dutton, 1994.

Crosby, Philip B. *Quality Is Free.* New York, McGraw-Hill Book Company, 1979.

Crosby, Philip B. *Running Things.* New York, McGraw-Hill Book Company, 1986.

Davidow, William H., and Bro. Uttal. *Total Customer Service.* New York, Harper & Row Publishers, Inc., 1989.

Deming, W. Edwards. *Out of the Crisis.* Cambridge, MA, Massachusetts Institute of Technology, Center for Advanced Engineering Study, 1982.

Department of Defense. CSDL-R-2161, *Findings of the U.S. Department of Defense Technology Assessment Team on Japanese Manufacturing Technology,* Final Report. June 1989.

Department of Defense. *Total Quality Management Guide,* vols. 1 and 2, Final Draft. Washington, DC, Feb. 15, 1990.

Drucker, Peter. *Management: Tasks, Responsibilities, and Practices.* New York: Harper & Row Publishers, Inc., 1974.

Drucker, Peter F. *The New Realities.* New York, Harper & Row Publishers, Inc., 1989.

Feigenbaum, Armand V. *Total Quality Control.* New York, McGraw-Hill Book Company, 1991.

Garvin, David. "How the Baldrige Award Really Works," *Harvard Business Review,* November–December 1991, p. 94.

Goldratt, Eliyahu M., and Jeff Cox. *The Goal*. Croton-on-Hudson, NY, North River Press, 1986.

Guaspari, John. *I Know It When I See It*. New York, AMACOM American Management Association, 1985.

Hammer, Michael, and James Champy. *Reengineering the Corporation*. New York, Harper Business, 1993.

Harrington, H. James. *Business Process Improvement*. New York, McGraw-Hill, Inc., 1991.

Harrington, H. James. *The Improvement Process*. New York, McGraw-Hill Book Company, 1987.

Hersey, Paul, and Kenneth H. Blanchard. *Management of Human Behavior: Utilizing Human Resources*. Englewood Cliffs, NJ, Prentice Hall, 1988.

Hurdeski, Michael. *Computer Integrated Manufacturing*. Blue Ridge Summit, PA, TAB Books, Inc., 1988.

Hutchins, Greg. *ISO 9000*. Essex Junction, VT, Oliver Wight Publications, 1993.

Imai, Masaaki. *Kaizen*. New York, Random House, 1986.

Institute for Defense Analysis. *The Role of Concurrent Engineering in Weapons System Acquisition*, IDA Report R-338, December 1988.

Ishikawa, Kaoru. *Guide to Quality Control*. Tokyo, Asian Productivity Organization, 1982, New York, UNIPUB.

Jellison, Jerald M. *Overcoming Resistance*. New York, Simon and Schuster, 1993.

Jones, James V. *Integrated Logistics Support Handbook*. Blue Ridge Summit, PA, TAB Books, Inc., 1987.

Juran, Joseph M. *Juran on Planning for Quality*. New York, The Free Press, 1988.

Juran, Joseph M., and Frank M. Gryna, Jr. *Quality Planning and Analysis*. New York, McGraw-Hill Book Company, 1986.

King, Bob. *Better Designs in Half the Time: Implementing QFD Quality Function Deployment in America*. Methuen, MA, GOAL/QPC.

Levitt, Theodore. *The Marketing Imagination*. New York, The Free Press, 1980.

Levitt, Theodore. *Thinking About Management*. New York, The Free Press, 1991.

Lubben, Richard T. *Just-in-Time Manufacturing*. New York, McGraw-Hill Book Company, 1988.

Mansir, Brian E., and Nicholas R. Schacht. *Continuous Improvement Process*, Report IR806R1. Bethesda, MD, Logistics Management Institute, August 1989.

McGill, Michael E. *American Business and the Quick Fix*. New York, Henry Holt & Company, 1988.

Nakajima, Sceiichi. *Total Productive Maintenance*. Cambridge, MA, Productivity Press, 1988.

Nelson, Bob. *1001 Ways To Reward Employees*. New York, Workman Publishing Company, Inc., 1994.

Office of Deputy Assistant Secretary of Defense for TQM. *Total Quality Management: A Guide for Implementation* (DoD Guide 5000.51G). Washington, DC, 1989.

Peters, Thomas J. *Thriving on Chaos*. New York, Alfred A. Knopf, Inc., 1987.

Peters, Thomas J., and Robert H. Waterman, Jr. *In Search of Excellence*. New York, Harper & Row, 1982.

Peters, Tom. *Liberation Management*. New York, Alfred A. Knopf, 1992.

Ross, Phillip J. *Taguchi Techniques for Quality Engineering*. New York, McGraw-Hill Book Company, 1988.

Saylor, James H. *TQM Field Manual*. New York, McGraw-Hill, Inc., 1992.

Scherkenbach, William W. *The Deming Route to Quality and Productivity*. Washington, DC, Cee Press Books, 1988.

Scholtes, Peter R. *The Team Handbook*. Madison, WI, Joiner Associates, Inc., 1988.

Schonberger, Richard J. *World Class Manufacturing*. New York, The Free Press, 1986.

Senge, Peter. *The Fifth Discipline: The Art and Practice of the Learning Organization*. New York, Doubleday, 1990.

Shores, A. Richard. *Survival of the Fittest*. Milwaukee, WI, ASQC Quality Press, 1988.

Soin, Sarv Singh. *Total Quality Control Essentials*. New York, McGraw-Hill, Inc., 1992.

Taguchi, Genichi. *Introduction to Quality Engineering*. Tokyo, Asian Productivity Organization, 1986. Dearborn, MI, America Supplier Institute, Inc., 1986. White Plains, NY, UNIPUB/Quality Resources, 1989.

Thomas, Brian. *Total Quality Training*. New York, McGraw-Hill, Inc., 1992.

Townsend, Patrick L. *Commit to Quality*. New York, John Wiley & Sons, 1986.

Tunks, Roger. *Fast Track To Quality*. New York, McGraw-Hill, Inc., 1992.

Wallace, Thomas E. *MRPII: Making It Happen*. Essex Junction, VT, Oliver Wright Limited Publications, Inc., 1985.

Walton, Mary. *Deming Management at Work*. New York, Perigee Books, 1990.

Glossary

Activities   The steps of a process.

Agenda   A plan for the conduct of a meeting.

Appraisal costs   Costs associated with inspecting the product to ensure that it meets the customer's needs and expectations.

Bar chart   A chart for comparing many events or items.

Benchmarking   A method of measuring your organization against those of recognized leaders or best of class.

Best of class   One of a group of similar organizations whose overall performance, effectiveness, efficiency, and adaptability is superior to all others.

Brainstorming   Technique that encourages collective thinking power of a group to create ideas.

Cause   The reason for action or condition.

Cause-and-effect analysis   Technique for helping a group examine underlying causes; fishbone.

Charts   A graphic picture of data that highlights important trends and significant relationships.

Checksheet   A list made to collect data.

Coach   Person who acts as a guide for organization development.

Collaborate   To work jointly with others.

Commitment   Personal resolve to do something.

Common cause   Normal variation in an established process.

Communication   Technique for exchanging information.

Competition   Anyone or anything competing for customer(s).

Computer-aided design   Automated system for assisting in design process.

Computer-aided engineering   Automated system for assisting in engineering process.

Computer-aided manufacturing   Automated system for assisting process design for manufacturing.

Computer integrated manufacturing   The integration of CAD/CAM for all design and manufacturing processes.

Computer systems   Items such as hardware, software, firmware, robotics, expert systems, and artificial intelligence.

Concurrent engineering   Systematic approach to the integrated, concurrent design of products and their related processes, including manufacture and support. This approach is intended to cause the developers, from the outset, to consider all elements of the product life cycle from conception, through disposal, including quality, cost, schedule, and user requirements.

Consensus   An agreement reached by everyone which all members can understand and support.

Continuous improvement   The never-ending pursuit of excellence.

Continuous improvement system   A disciplined methodology to achieve the goal of commitment to excellence by continually improving all processes.

Control chart   A chart that shows process performance in relation to control limits.

Cooperate   To act together with others.

Corrective action   An action to correct an unwanted condition.

Cost of poor quality   Term for techniques that focus on minimizing the cost of nonconformance.

Cost of quality   Term for technique used to identify cost of conformance and nonconformance, which involves such factors as prevention, appraisal, internal failure, and external failure.

Criteria   A standard on which a decision can be based.

Culture   A prevailing pattern of activities, interactions, norms, sentiments, beliefs, attitudes, values, and products in an organization. The shared experience of a group.

Customer   Everyone affected by the product and/or service. The customer can be the ultimate user of the product and/or service, known as an external customer. Or the customer can be the next person or process in the organization, known as an internal customer.

Customer-driven   Process in which the customer or customer's voice is the primary focus and in which the customer leads the way. Customer satisfaction becomes the focus of all efforts. This goal provides the constancy of purpose vital to success.

Customer satisfaction   In Total Quality Management, quality.

Customer/supplier analysis   Techniques that provide insight into the customer's needs and expectations and involve an organization's suppliers in the development of an organization's requirements and their supplier's conformance to them.

Cycle time   The time from the beginning of a process to the end of a process.

Data Information or a set of facts.

Data statistical analysis Tools for collecting, sorting, charting, and analyzing data to make decisions.

Decision making The process of making a selection.

Defect Any state of nonconformance to requirements.

Deliverable The output of a process provided to a customer. The deliverable can be a product or service or a combination of a product and services.

Design of experiments Traditionally, an experimental tool used to establish both parametric relationships and a product/process model in the early (applied research) stages of the design process.

Design phases The three phases of the design of a product or process are, according to Taguchi: systems design, parameter design, and tolerance design.

Detailed process diagram A flowchart, consisting of symbols and words, that completely describes a process.

Detection Identification of nonconformance after the fact.

Deviation Any nonconformance to a standard or requirement.

Disciplined continuous improvement methodology The continuous improvement system.

Driving forces Those forces that are pushing toward the achievement of a goal.

Effect A problem or defect that occurs on the specific job to which each group or team is assigned.

Effectiveness A characteristic used to describe a process in which the process output conforms to requirements.

Efficiency A characteristic used to describe a process that produces the required output at a perceived minimum cost.

Empowerment The power of people to do whatever is necessary to do the job and improve the system within their defined authority, responsibility, and resources.

External customer The ultimate user of the product and/or service.

Extrinsic rewards Rewards given by other people.

Facilitator One who assists the team in developing teamwork and applying the TQM tools and techniques.

Fishbone See *Cause-and-Effect Analysis.*

Flow diagram A drawing combined with words used for defining a process. This tool provides an indication of problem areas, unnecessary loops and complexity, non-value-added tasks, and areas where simplification of a process is possible.

Focus setting Technique used to focus on a specific outcome.

Force-field analysis   Technique that helps a group describe the forces at work in a given situation.

Functional organization   An organization responsible for a major organizational function such as marketing, sales, design, manufacturing, and distribution.

Functional team   A team consisting of representatives from only one functional area.

GANTT chart   A graphic representation of the project network placed in the context of a calendar to relate tasks to real time.

Goal   The specific desired outcome.

Guideline   A suggested practice that is not mandatory in programs intended to comply with a standard.

Hierarchical nature of a process   The various levels of a process.

Histogram   A chart that shows frequency of data in column form.

Histogram (project)   A graphic barchart, comparing the resource needs of various project tasks with one another and with available resources over time.

House of Quality   Quality functional deployment (QFD) planning chart.

Human resources   The people in an organization.

Improvement methodology   A method for making improvements in an organization.

In control   A process within the upper and lower limits.

Individual involvement   Involvement of each person in the output of an organization.

Information system   Automated systems used throughout the organization to review, analyze, and take corrective action.

Input   What is needed to do the job.

Input/output analysis   Technique for identifying interdependency problems.

Institutionalize   To make an integral part of the organization's way of life.

Internal customer   Next person or process in the organization.

Intrinsic rewards   Rewards that are an integral part of the system. These rewards are within the individual person.

Just-in-time   Method and philosophy of having the right material just in time to be used in an operation, which eliminates transactions and uses the whole person.

Lead team   A team that oversees several other teams.

Leadership   Guidance of a group of people to accomplish a common goal.

Life cycle cost   The total cost of system or item over its full life. This includes the cost of acquisition, ownership, and disposal.

Line chart   A chart that describes and compares quantifiable information.

Listening   Technique for receiving and understanding information.

Loss function   A function that examines the costs associated with any variation from the target value of a quality characteristic.

Lower control limit   The lower control limit of the process, usually minus 3 sigma of the statistic.

Maintenance   The repair of an item.

Malcolm Baldrige National Quality Award   An annual award created by public law to recognize U.S. companies that excel in quality achievement and quality management to help improve quality and productivity.

Management   The leadership of an organization.

Management functions   Functions that include planning, organizing, staffing, directing, controlling, and coordinating.

Management involves optimizing resources   This process involves getting the most out of both technology and people. The target is on managing the project and leading the people to a deliverable that totally satisfies the customer.

Manufacturing resource planning II   System for planning and controlling a manufacturing company's operation.

Matrix organization   An organization where resources are shared between both functional management and project management.

Mean   The average of a group of data.

Mean time between failures (MTBF)   The average time between successive failures of a given product.

Measurement   The act or process of measuring to compare results to requirements. A quantitative estimate of performance.

Meeting   Technique of bringing a group together to work for a common goal.

Mentor   A person assigned as management interface support for a team.

Metric   Meaningful measures that focus on total customer satisfaction and target improvement action.

Mission   The intended result. The basic organizational view of the role and function of the organization in satisfying customers' expectations today and in the future.

Mistake proofing   Technique for avoiding simple, human error at work; poka-yoke.

Multifunctional team   A team consisting of representatives from more than one function.

Noise   A factor that disturbs the function of a process or function.

Nominal group technique   Technique similar to brainstorming that provides structured discussion and decision making.

Non-value-added Term used to describe a process, activity, or task that does not provide any value to the product.

Out-of-control process A process for which the outcome is unpredictable.

Output The results of the job.

Owner The person who can change the process without further approval.

Ownership The power to have control over.

Paradigm An expected pattern or view.

Parameter (or robust) design This stage focuses on making the product performance (or process output) insensitive to variation by moving towards the best target values of quality characteristics.

Parametric design The design phase where the sensitivity to noise is reduced.

Pareto's principle The principle that a large percentage of the results are caused by a small percentage of the causes; for instance, 80 percent of results are caused by 20 percent of causes.

People involvement Individual and group activities involving people.

Performance A term used to describe both the work product itself and a general process characteristic. The broad performance characteristics that are of interest to management are quality (effectiveness), cost (efficiency), and schedule. Performance is the highly effective common measurement that links the quality of the work product to efficiency and productivity.

Pie chart A chart in circular form that is divided to show the relationship between items and the whole.

Plan A specified course of action designed to attain a stated objective.

Poka-yoke See *Mistake proofing.*

Policy A statement of principles and beliefs, or a settled course, adopted to guide the overall management of affairs in support of a stated aim or goal. It is mostly related to fundamental conduct and usually defines a general framework within which other business and management actions are carried out.

Population A complete collection of items (product observations, data) about certain characteristics of which conclusions and decisions are to be made for purposes of process assessment and quality improvement.

Potential product Anything that can be used to attract and hold customers beyond the augmented product.

Presentation Tool for providing information, gaining approval, or requesting action.

Prevention A future-oriented approach to quality management that achieves quality improvement through corrective action on the process.

Prevention Costs Costs associated with actions taken to plan the product or process to ensure that defects do not occur.

Problem   A question or situation proposed for solution. The result of not conforming to requirements or, in other words, a potential task resulting from the existence of defects.

Process   A series of activities that takes an input, modifies the input (work takes place and value is added), and produces an output.

Process analysis   Tool used to improve the process and reduce process time by eliminating non-value-added activities and/or simplifying the process.

Process capability   Long-term performance level after the process has been brought under control.

Process control   In statistics, the set of activities employed to detect and remove special causes of variation in order to maintain or restore stability.

Process design   The development of a process.

Process diagram   A tool for defining a process.

Process improvement   The set of activities employed to detect and remove common causes of variation in order to improve process capability. Process improvement leads to quality improvement.

Process improvement team   A team of associates with representative skills and functions to work specific process(es).

Process management   Management approach comprising quality management and process optimization.

Process optimization   The major aspect of process management that concerns itself with the efficiency and productivity of the process, that is, economic factors.

Process owner   A designated person within the process who has authority to manage the process and is responsible for its overall performance.

Process performance   A measure of how effectively and efficiently a process satisfies customer requirements.

Process review   An objective assessment of how well the methodology has been applied to your process. Emphasizes the potential for long-term process results rather than the actual results achieved.

Product   An output of a process provided to a customer (internal/external), including goods, systems, equipment, hardware, software, services, and information.

Product design   The development of the product.

Production process   The manufacturing of the product.

Productivity   The value added by the process divided by the value of the labor and capital consumed.

Project   Any series of activities that has a specific end objective. Almost all activities in an organization can be defined in terms of a project.

Project management   The process of planning and controlling the coordinated work of a project team to produce a product, or deliverable, for the customer within defined cost, time, and quality constraints.

Project schedule   A plan for carrying out the tasks in sequence on the basis of their interdependency.

Project objective   A narrative statement of the project deliverable, its basic dimensions, and a description of how the deliverable will improve quality and customer satisfaction.

Pull   The customer drives the deliverable from the marketplace.

Push   The marketplace creates the demand of the deliverable for the customer.

Quality   In Total Quality Management, total customer satisfaction.

Quality (DoD)   Conformance to a set of customer requirements that, if met, results in a product or service that is fit for its intended use.

Quality (product)   Conformance to requirements.

Quality (Taguchi)   The (measure of degree of) loss a product causes after being shipped, other than any losses caused by its intrinsic functions.

Quality function deployment   A disciplined approach for transforming customer requirements (the voice of the customer) into product development requirements at each phase of product development.

Quality improvement team   A group of individuals charged with the task of planning and implementing quality improvement.

Quantitative methods   Use of measurements.

Range   The difference between the maximum and the minimum value of data.

Recognition   Special attention paid to an individual or group.

Reengineer   The fundamental rethinking and radical redesign of business processes to achieve dramatic improvements in critical contemporary measures of performance, such as cost, quality, service, and speed. (For more information, see *Reengineering the Corporation* by Michael Hammer and James Champy.)

Reliability   The probability that an item will perform its intended function for a specified interval under stated conditions.

Requirement   A formal statement of a need and the expected manner in which it is to be met.

Requirements   Expectations for a product or service. The "it" in "Do it right the first time." Specific and measurable customer needs with an associated performance standard.

Restraining forces   Forces that keep a situation from improving.

Reward   Recompense, either external or internal. External rewards are controlled by other people; they are pay, promotion, and benefits. Internal rewards are part of the task or individual; they are items like challenge, feeling of accomplishment, feeling of belonging, and sense of pride.

Risk management   A technique for continually assessing the risk in each task of the project not only in terms of time and cost but also in technical feasibility of the task.

Robust design   Design of a product so minimal quality losses are incurred.

Root cause   Underlying reason for nonconformance within a process. When the root cause is removed or corrected, the nonconformance will be eliminated.

Rules of conduct   Rules that provide guidance for the team's conduct.

Sample   A finite number of items taken from a population.

Sampling   The collection of some, not all, of the data.

Scatter chart   A chart that depicts the relationship between two or more factors.

Scope of work   A broad narrative discussion of the project objective, tasks, and project network, placing limits on the project and defining how the work will progress.

Selection grid   Tool for comparing each problem, opportunity, or alternative against all others.

Selection matrix   Technique for rating problems, opportunities, or alternatives based on specific criteria.

Service   A range of customer satisfiers.

Seven basic tools of quality   The seven tools are Pareto charts, cause-and-effect diagrams, stratification, checksheets, histograms, scatter diagrams, and control charts.

Simulation   Technique of observing and manipulating an artificial mechanism (model) that represents a real-world process that, for technical or economical reasons, is not suitable or available for direct experimentation.

Solution   The answer to a problem.

Sorting   Arranging information into some order, such as into classes or categories.

Special/assignable causes   Abnormal causes of variation in the process.

Specification   A document containing a detailed description or enumeration of particulars. Formal description of a work product and the intended manner of providing it. (The provider's view of the work product.)

Stages of team development   The phases through which every team goes, from orientation to dissatisfaction, resolution, and production.

Standard deviation   A parameter describing the spread of the process output. The positive square root of the variance.

Statistic   Any parameter that can be determined on the basis of the quantitative characteristics of a sample.

Statistical control   Term used to describe a process from which all special causes of variation have been removed and only common causes remain. Such a process is also said to be stable.

Statistical estimation   The analysis of a sample parameter in order to predict the values of the corresponding population parameter.

Statistical methods   The application of the theory of probability to problems of variation.

Statistical process control (SPC)   Statistical tool for monitoring and controlling a process to maintain and possibly improve quality.

Statistics   The branch of applied mathematics that describes and analyzes empirical observations for the purpose of predicting certain events in order to make decisions in the face of uncertainty.

Steering group   An executive-level steering committee.

Strategy   A broad course of action, chosen from a number of alternatives, to accomplish a stated goal in the face of uncertainty.

Stratification   Arranging data into classes.

Subprocesses   The internal processes that make up a process.

Supplier   An individual, organization, or firm that provides inputs to a process. The supplier can be internal or external to a company, firm, or organization.

Supplier/customer analysis   Technique used to obtain and exchange information for conveying an organization's needs and requirements to suppliers and mutually determining needs and expectations of the customers.

Support system   A system within the organization that guides it through the TQM process.

Synergy   A team of people working together in a cooperative effort.

System   Many processes combined to accomplish a specific function.

System improvement   A method that focuses on the development or redesign of systems.

Systems (parts) or concept design   This phase arrives at the design architecture (size, shape, materials, number of parts, etc.) by looking at the best available technology. Note that this phase is commonly known as the parts design phase in American terminology.

Taguchi approach   Techniques for reducing variation of product or process performance to minimize loss.

Task   One of a number of actions required to complete an activity.

Task list   The list of individual tasks determined necessary to produce all the components of the work, broken down into categories and arranged in a structure.

Team   Group of people working together toward a common goal.

Teamwork   Shared responsibility for the completion of a common task or problem.

Tolerance design   A stage of design that focuses on setting tight tolerances to reduce variation in performance. Because this phase is the one most responsible for adding costs, an essential goal is a reduction in the need for setting tight tolerances by successfully producing robust products and processes in the parameter design phase.

Top-down process diagram   A chart of the major steps and substeps in the process.

Top-level process diagram   A diagram of the entire, overall process.

Total   Term used to describe the involvement of everyone and everything in a continuous improvement effort. Everyone is committed to "one" common organizational purpose as expressed in the vision and mission. They are also empowered to act to make it a reality. Besides people, everything in the organization, including systems, processes, activities, tasks, equipment, and information, must be aligned toward the same purpose.

Total production maintenance   System for involving the total organization in maintenance activities.

Total Quality Management (TQM)   A leadership and management philosophy and guiding principles stressing continuous improvement through people involvement and quantitative methods focusing on total customer satisfaction.

Total Quality Management action planning   The specific road map for the overall TQM effort that establishes a clear focus for the organization.

TQM environment   An internal organizational environment of openness, honesty, trust, communication, involvement, ownership, pride of workmanship, individuality, innovation, creativity, and personal commitment to be the "best."

TQM philosophy   The overall, general concepts for a continuously improving organization.

TQM principles   The essential fundamental rules required to achieve victory.

TQM process   The process that transforms all the inputs into the organization into a product and/or service that satisfies the customer.

TQM umbrella   The integration of all the fundamental management techniques, existing improvement efforts, and technical tools under a disciplined approach focused on continuous improvement.

Trainers   People in an organization who provide skills training.

Training   The teaching of skills, knowledge, and attitude to accomplish actions.

Upper control limit   The upper control limit of a process usually plus 3 sigma of the statistic.

Values   The principles that guide the conduct of an organization.

Variable   A data item that takes on values within some range with a certain frequency or pattern.

Variance   In quality management terminology, any nonconformance to specifications. In statistics, it is the square of the standard deviation.

VICTORY-C TQM model   A systematic, integrated, consistent, organization-wide model consisting of all the elements required for VICTORY focused on total customer satisfaction.

VICTORY elements   The following elements are required for VICTORY focused on total customer satisfaction; customer focus, leadership, vision, involvement of everyone and everything, continuous improvement of all systems and processes, training and education, ownership, reward and recognition, and yearning for success.

Vision   Where the organization wants to go.

Vote   Technique to determine majority opinion.

Index

ABOUT THE AUTHOR

James H. Saylor is founder of The Business Coach, a consulting firm focusing on helping organizations achieve their specific VICTORY. He has assisted many small, medium, and large organizations to success using Total Quality Management. He has over 28 years of experience performing organizational development, total quality, project management, logistics, and training for several public and private organizations. He is the author of *TQM Field Manual* and coauthor of *Customer-Driven Project Management*, both published by McGraw-Hill.